PRACTICAL WASTEWATER TREATMENT

PRACTICAL WASTEWATER TREATMENT

David L. Russell, PE

Global Environmental Operations, Inc.
Lilburn, Georgia

A John Wiley & Sons, Inc., Publication

Library of Congress Cataloging-in-Publication Data:

Russell, David L. (David Lloyd), 1943-
  Practical wastewater treatment / David L. Russell.
    p. cm.
  Includes index.
  ISBN-13: 978-0-471-78044-1 (cloth)
  ISBN-10: 0-471-78044-8 (cloth)
1. Water treatment plants. 2. Sewage–Purification. I. Title.

TD434.R87 2006
628.1'683–dc22   2006010287

Printed in the United States of America

10 9 8 7 6 5 4 3 2 1

To Marianne for her invaluable support, and Laura and Jennifer for being themselves

CONTENTS

PREFACE

Over the years I have been fortunate enough to meet some very interesting people and have some fantastic experiences in the environmental field. In 1998, Nina Webber, the then Educational Director for the American Institute of Chemical Engineers approached me to teach a course in wastewater treatment because an existing instructor balked at the assignment of going on to teach in Mexico. This book was developed from that teaching assignment.

This is a teaching tool for the chemical and environmental engineering professionals. It is not designed to be a textbook or primer for those entering this profession because it lacks adequate development of theory for that purpose and relies upon plant experience and a mastery of essential engineering fundamentals for many of the subjects. This book is more a cross between a chemical engineering handbook and a refresher tool for the plant engineer who suddenly finds himself or herself having to learn to water and wastewater treatment and does not know where to start. I hope that it serves that purpose.

The theoretical development generally tends to be sparse except in the area of biological wastewater treatment and some elements of hydraulics. I have also placed a good bit of emphasis on the development of biological modeling of wastewater treatment plants because I firmly believe that it is the best way to design facilities, and it is the wave of the future. I have, through my own work, found out that most wastewater treatment plants designed by municipal codes are between 30% and 50% overdesigned, and when the consultant applies a standard allowance for growth that often means that the plant is 100% or more overdesigned and wasteful of precious municipal resources and money. The design of a system with that much additional capacity leads to sloppy operation and poor control. It also leads to a perception that wastewater plant operators do not need to understand the biological processes, and that they are little more than mechanics.

Finally, I have included some design hints and practical experience where it may be helpful. The focus has been to provide a framework of useful tools and helpful aids where they can be found, including links to the World Wide Web, and various other textbooks where they treat specific subjects.

I have taken some pains to assemble various Web sources and references, including helpful papers and articles and even computer programs on to a disk, which was originally supplied as a supplement to the course. The disk is available from me for a modest fee.

Questions, comments, flames, and other stuff should be directed to my attention via my e-mail address: dlr@mindspring.com

Dave Russell
March, 2006

1

INTRODUCTION

INTRODUCTION

This course is almost all about water and its treatment. In it we will look at all the phases of the water environment, the types and characteristics, and contaminants. We will also discuss briefly various elements of hydrology and hydraulics, but only in enough detail, to permit you to get familiar with each subject and recognize some pitfalls and common mistakes chemical engineers make when they deal with the water environment.

We will also look at the elements of biological treatment in some depth because it is important to know what the limits of biological systems are, and more important, where are they applicable and inapplicable. We will look at some of the pitfalls inherent in the measurement systems we use and even take a

Practical Wastewater Treatment, by David L. Russell
Copyright © 2006 John Wiley & Sons, Inc.

brief look at the accuracy of our measurements so that we may adequately characterize the materials we discharge to the regulatory community.

WATER COMPOSITION

Water is composed of two parts hydrogen and one part oxygen. It is not the materials of the water but the contaminants in it that make it important. If we look at a chemical reaction, we would be happy with a yield of 99.95% purity. However, for water this level of impurity is unacceptable. We are dealing with a substance where the levels of contaminants that we often consider insignificant can spoil the quality or use of the product.

Examples of the forgoing are things like salinity or dissolved NaCl, which if present in levels of 500 ppm or higher render water marginally potable. Another shining example: the presence of as little as 1 ppm of lead, 10 ppm of nitrate, 10 ppm of sewage solids, or 5 ppm of the right detergent will render the water unusable.

PURE WATER

Characteristics:

Formula: H_2O

Dissolved Gases: The most important is oxygen and the second most important is nitrogen

Solubility of Gases in Water: Solubility (See Table 1.1).

The concentration of oxygen in water at any pressure is given by:

$$\text{Ln } C = -139.34411 + (1.575701 \times 10^{+5}/T) - (6.642308 \times 10^{+7}/T^2)$$
$$+ (1.243800 \times 10^{+10}/T^3) - (8.621949 \times 10^{+11}/T^4)$$
$$- \mathbf{Chl}[\{3.1929 \times 10^{+2}\} - \{1.9428 \times 10^{+1}/T\} + \{3.8673 \times 10^{+3}/T^2\}]$$

where **Chl** is the chlorinity measured in grams/kilogram and is defined as:

$$\text{Chlorinity} = \text{Salinity}/1.80655$$

Salinity is approximately equal to total solids in water after carbonates have been converted to oxides and after all bromide and iodide have been replaced by chloride.[1]

Nitrogen is soluble in water too, but the presence of nitrogen in the gaseous or N_2 form is essentially inert. Principal forms of nitrogen in water are ammonia, nitrate, and nitrite. The only time one has to worry about the solubility of nitrogen or other gases in water is when one is designing a pressure flotation system.

[1]See Standard Methods Oxygen Dissolved Method 4500-O/Azide Modification.

TABLE 1.1   Solubility of Oxygen in Water Exposed to Water-Saturated Air at Atmospheric Pressure (101.3 kPa)1

	Oxygen Solubility in mg/l					
	Temperature					
Chlorinity:	0	5	10	15	20	25
0	14.621	13.728	12.888	12.097	11.355	10.657
1	14.216	13.356	12.545	11.783	11.066	10.392
2	13.829	13.000	12.218	11.483	10.790	10.139
3	13.460	12.660	11.906	11.195	10.526	9.897
4	13.107	12.335	11.607	10.920	10.273	9.664
5	12.770	12.024	11.320	10.656	10.031	9.441
6	12.447	11.727	11.046	10.404	9.799	9.228
7	12.139	11.442	10.783	10.162	9.576	9.023
8	11.843	11.169	10.531	9.930	9.362	8.826
9	11.559	10.907	10.290	9.707	9.156	8.636
10	11.288	10.656	10.058	9.493	8.959	8.454
11	11.027	10.415	9.835	9.287	8.769	8.279
12	10.777	10.183	9.621	9.089	8.586	8.111
13	10.537	9.961	9.416	8.899	8.411	7.949
14	10.306	9.747	9.218	8.716	8.242	7.792
15	10.084	9.541	9.027	8.540	8.079	7.642
16	9.870	9.344	8.844	8.370	7.922	7.496
17	9.665	9.153	8.667	8.207	7.770	7.356
18	9.467	8.969	8.497	8.049	7.624	7.221
19	9.276	8.792	8.333	7.896	7.483	7.090
20	9.092	8.621	8.174	7.749	7.346	6.964
21	8.915	8.456	8.021	7.607	7.214	6.842
22	8.743	8.297	7.873	7.470	7.087	6.723
23	8.578	8.143	7.730	7.337	6.963	6.609
24	8.418	7.994	7.591	7.208	6.844	6.498
25	8.263	7.850	7.457	7.083	6.728	6.390
26	8.113	7.711	7.327	6.962	6.615	6.285
27	7.968	7.575	7.201	6.845	6.506	6.184
28	7.827	7.444	7.079	6.731	6.400	6.085
29	7.691	7.317	6.961	6.621	6.297	5.990
30	7.559	7.194	6.845	6.513	6.197	5.896
31	7.430	7.073	6.733	6.409	6.100	5.806
32	7.305	6.957	6.624	6.307	6.005	5.717
33	7.183	6.843	6.518	6.208	5.912	5.631
34	7.065	6.732	6.415	6.111	5.822	5.546
35	6.950	6.624	6.314	6.017	5.734	5.464
36	6.837	6.519	6.215	5.925	5.648	5.384
37	6.727	6.416	6.119	5.835	5.564	5.305
38	6.620	6.316	6.025	5.747	5.481	5.228
39	6.515	6.217	5.932	5.660	5.400	5.152
40	6.412	6.121	5.842	5.576	5.321	5.078
41	6.312	6.026	5.753	5.493	5.243	5.005

TABLE 1.1   (*Continued*)

Oxygen Solubility in mg/l						
	Temperature					
Chlorinity:	0	5	10	15	20	25
42	6.213	5.934	5.667	5.411	5.167	4.933
43	6.116	5.843	5.581	5.331	5.091	4.862
44	6.021	5.753	5.497	5.252	5.017	4.793
45	5.927	5.665	5.414	5.174	4.944	4.724
46	5.835	5.578	5.333	5.097	4.872	4.656
47	5.744	5.493	5.252	5.021	4.801	4.589
48	5.654	5.408	5.172	4.947	4.730	4.523
49	5.565	5.324	5.094	4.872	4.660	4.457
50	5.477	5.242	5.016	4.799	4.591	4.392

Henry's law gives us some idea of the solubility of other gases.

Stating the pressure–concentration ratio as an equation and using the usual modern symbol for the Henry's law constant on a concentration basis give the following form of Henry's law:

$$p = K'_c c$$

In this form p is the partial pressure of the gas, c is its molar concentration, and K'_c is the Henry's law constant on the molar concentration scale. Henry's law is found to be an accurate description of the behavior of gases dissolved in liquids when concentrations and partial pressures are reasonably low. As the concentrations and partial pressures increase, deviations from Henry's law become noticeable. This behavior is very similar to the behavior of gases, which are found to deviate from the ideal gas law as pressures increase and temperatures decrease. For this reason, solutions that obey Henry's law are sometimes called ideal dilute solutions.

Values of the Henry's law constants for many gases in many different solvents have been measured. Table 1.2 gives a few selected values of the Henry's law constants for gases dissolved in water.

Values in this table are calculated from tables of molar thermodynamic properties of pure substances and aqueous solutes.

The inverse of the Henry's law constant, multiplied by the partial pressure of the gas above the solution, is the molar solubility of the gas. Thus oxygen at one atmosphere would have a molar solubility of $(1/756.7)\,mol/dm^3$ or $1.32\,mmol/dm^3$.

The following examples will help in understanding this concept.

TABLE 1.2   Molar Henry's Law Constants for Aqueous Solutions at 25°C

Gas	Constant $(Pa/(mol/dm^3))$	Constant $(atm/(mol/dm^3))$
He	$282.7 \times 10^{+6}$	2865.0
O_2	$74.68 \times 10^{+6}$	756.7
N_2	$155 \times 10^{+6}$	1600.0
H_2	$121.2 \times 10^{+6}$	1228.0
CO_2	$2.937 \times 10^{+6}$	29.76
NH_3	$5.69 \times 10^{+6}$	56.9

Example 1: The amount of oxygen dissolved in air-saturated water under normal atmospheric conditions at 25°C can be calculated as follows. Normal atmospheric condition is 20.948 mol% oxygen, which makes the partial pressure of oxygen 0.20948 atm or 20.67 kPa. Using Henry's law, the concentration of oxygen is $0.20948\,atm/(756.7\,atm/(mol/dm^3))$, which is $2.768 \times 10^{-4}\,mol/dm^3$ or $0.2768\,mmol/dm^3$, given the weight of 32 g/mol that comes out to be $0.0000088576\,g/dm^3$ or about 8.85 mg/l, which is to be compared with the tabular value of 8.23 mg/l from Table 1.2.

Example 2: If we want to run a dissolved air flotation system at 50 psig (115.23 ft of water pressure or 3.4473785 bar) for the pressure for flotation, how much nitrogen and oxygen will be produced when we release the pressure back to atmospheric?

The density of water is about $1\,kg/dm^3$ or $1000\,kg/m^3$. The pressure is approximately equal to a column of water 35.15344 m high. A column of water 35.15 m high would exert a pressure of $35153.44\,kg/m^2$ of its base, which converts to 344.73748 kPa pressure. The total system pressure is atmospheric pressure plus compression or 101.325 kPa + 344.7375 kPa or a total of 446.0625 kPa. (This is equivalent to $446.0625/101.325 = 4.4023\,atm$.) The pressure change of 3.4023 atm (4.4023 atm − 1 atm) will produce a concentration change of $3.4023/1600 = 0.0021264375\,mol/dm^3$.[2] (The pressure change of 344.738 kPa will cause a concentration change of $2.12644\,mmol/dm^3$). For each gallon of water the amount of nitrogen generated is $3.785 \times 2.12644\,mmol = 8.418\,mmol$ or $0.00666\,ft^3$ of nitrogen per gallon, or about 189 ml of nitrogen. For oxygen, the change is about $4.496\,mmol/dm^3$ or about 100.7 ml of O_2 per liter or about 382 ml per cubic foot. The total volume for flotation is about 571 ml of gas per cubic foot.

---

[2]Note that the difference in constants does cause some differences in the concentration and volume in the second and third decimal places and beyond.

The value of the Henry's law constant is temperature dependent. The value generally increases with increasing temperature. As a consequence, the solubility of gases generally decreases with increasing temperature. One example of this can be seen when water is heated on a stove. The gas bubbles appearing on the sides of the pan well below the boiling point of water are bubbles of air, which evolve due to the lowered solubility from hot water. The addition of boiled or distilled water to a fish tank will cause the fish to die of suffocation unless the water has been allowed to re-aerate before addition.

A very complete listing of many Henry's law constants can be found at http://www.mpch-mainz.mpg.de/~sander/res/henry.html#3. The file is in Adobe Acrobat and Zip formats. A computer program for calculating Henry's law coefficients can be found on the World Wide Web at http://www.syrres.com/esc/est_soft.htm. A specific value for a Henry's coefficient determined by one researcher may disagree with the same coefficient determined by another researcher by an order of magnitude.

If you have one value for a Henry's coefficient at a given set of conditions, (atm m^3/mol) it can be transformed to another set of conditions by the equation:

$$H_{TS} = H_R \times \exp[-\Delta H_{V,TS}/R_c(1/T_S - 1/T_R)]$$

where H_{TS} is the coefficient at temperature T_S, and T_R is the reference temperature in K (kelvin). The term $\Delta H_{V,TS}$ is the enthalpy of vaporization at T_S in units of cal/mol, and R_c is the gas constant, which has units of 1.9872 cal/mol K. The enthalpy can be obtained either from steam tables for water or chemical engineering tables for other fluids, or by using an alternative procedure for estimating the enthalpy of vaporization from the USEPA Web site: http://www.epa.gov/athane/learn2model/part-two/onsite/esthenry-background.htm.

The study of Henry's law has been of interest to the chemical engineering community for a long time. However, when the problems of benzene, toluene, and MTBE in groundwater were encountered, the subject regained renewed interest from the environmental community because of the use of Henry's law in strippers designed to remove the benzene, toluene, ethylbenzene, xylene, and MTBE resulting from a gasoline spill or tank release. MTBE cannot be removed effectively by stripping alone. Henry's coefficients may not really be considered a constant but will vary with temperature and pressure.

SALTS AND IONS IN WATER

There are a variety of salts in water. The most abundant salt in water is sodium chloride or NaCl. Table 1.3 shows the approximate concentration of the principal dissolved elements in seawater.

TABLE 1.3 Approximate Concentration of Principal Dissolved Elements in Seawater

Element	Concentration Coefficient (mg/l)	Exponent (10)	Element	Concentration Coefficient (mg/l)	Exponent (10)
Oxygen	8.57	5	Potassium	3.8	2
Hydrogen	1.08	5	Bromine	2.8	1
Chlorine	1.9	4	Strontium	8.1	0
Sodium	1.05	4	Boron	4.6	0
Magnesium	1.35	3	Silicon	3	0
Sulfur	8.85	2	Fluorine	3	0
Calcium	4	2	Argon	6	−1

Source: Handbook of Chemistry and Physics – 66 ed.

Later on, we will see that sodium salt is the most important salt in water, while calcium and magnesium salts are the most abundant in freshwater, and the interactions between carbon dioxide and lime stone (calcium carbonate and magnesium carbonate formations) also play a significant role in water and water treatment.

PRINCIPAL CONTAMINANTS AND IONS IN WATER AND MEASUREMENT METHODS

If we are going to consider the concentrations of chemicals in water, we must also have some knowledge about the way in which the chemicals are measured. This is not a text on analytical chemistry but merely a brief mention of some of the methods of detecting the most common compounds dissolved in water.

In analytical industry there are two principal references on methods. The first and oldest one is *Standard Methods for the Examination of Water and Wastewater*, published by the American Water Works Association, the Water Environment Federation, and the American Public Health Association. The second one has become important not only because of its publisher: *SW-846, Test Methods for Evaluating Solid Waste, Physical/Chemical Methods*, which was originally published by the United States Environmental Protection Agency (USEPA) Office of Solid Waste Research, principally for hazardous waste analyses, but also for many of the methods that are applicable to groundwater and wastewater. It has also become a de facto standard in the United States and elsewhere because of the many references in EPA-issued permits to the manual. The manual can be viewed and downloaded at http://www.epa.gov/epaoswer/hazwaste/test/main.htm. However, it often does not include as thorough an explanation of the methods

TABLE 1.4 Analytical Methods Used for Compounds in Water

Element	Measurement Method	Element n	Measurement Method
Aluminum	Flame ionization	Carbonate (CO_3)	Calculation
Antimony	Flame ionization	Chloride (Cl)	Gravimetric
Arsenic	Flame ionization	Cyanide (CN)	Colorimetric
Calcium	Flame ionization	Fluoride (F)	Gravimetric
Chromium	Flame ionization	Hydronium (OH)	pH
Copper	Flame ionization	Hypochlorite ($HClO_2$)	pH
Hydrogen	pH	Hypochlorous (ClO_2)	pH
Iron	Flame ionization	Nitrate (NO_3)	Colorimetric
Lead	Flame ionization	Nitrite (NO_2)	Colorimetric
Magnesium	Flame ionization	Sulfate (SO_4)	Colorimetric
Manganese	Flame ionization	Sulfite (S)	Colorimetric
Mercury	Flame ionization	**OTHER**	
Potassium	Flame ionization	Alkalinity	Colorimetric
Silica	Flame ionization	Total org. carbon	Gravimetric
Silver	Flame ionization	Diss. O_2	Azide titr or probe
Sodium	Flame ionization	Org. nitrogen	Kjelldahl
Zinc	Flame ionization	Chem O_2 Demand	Digestion/titration
Ammonia	Kjelldahl or Nesslerization	Biochemical. O_2 Demand	Difference in oxygen uptake
Bicarbonate (HCO_3)	Calculation		

and the procedures as *Standard Methods*, and any good laboratory will have both. The test methods are slightly different, and in some cases, especially where more conventional parameters are involved, *SW-846* is silent.

SOURCES OF WATER

Groundwater

There are several sources for water. Groundwater serves the majority of the small communities in the United States, and elsewhere in the world. It is a source of drinking water. Groundwater is characterized by natural minerals in moderate to low concentrations. It is necessary to mention groundwater because it is most commonly ignored (being out of sight) and we do not often think about the need to protect the groundwater.

Flow regimens in groundwater are linear, and flow through porous media is analogous to heat transfer through solid in a solid medium. The overall equations used to calculate flow regimens are the Darcy equations, and they are laminar flow.

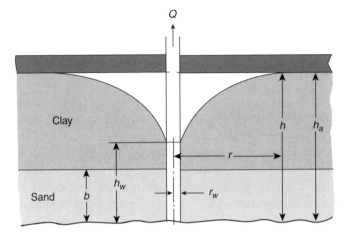

r_w = Radius of well

K = Coefficient of permeability

h_w = Height of water in the well being pumped

b = Thickness of aquifer

h_a = Hydraulic head at nonpumping conditions

Q = Pumping rate

$$Q = Z\pi rbK \frac{dh}{dr}$$

$$h - h_w = \frac{Q}{2\pi Kb} \ln \frac{r}{r_w} \quad \text{or} \quad h - h_w = \frac{Q}{2\pi T} \ln \frac{r}{r_w}$$

FIGURE 1.1   Basic groundwater flow equations.

The basic groundwater flow equation is shown below. The equation is in SI units.

$$Q(\text{flow}) = \pi^* K[(H^2 - h^2)]/\log_e(\{D/2\}/\{d/2\})$$

where the characters apply to the drawing shown in Figure 1.1.

The constant K is known as the permeability coefficient and it is given in velocity units of gallons per day per square feet or CuM/D/SqM (M/D) units.

The model used above is the simplest in an extremely complex set of possible combinations because the ground is not a homogeneous medium. I am raising this point because once contaminated, the groundwater is difficult, if not nearly impossible, to decontaminate. Groundwater protection must be a plant-wide priority.

Surface Water

Surface water has its origins in groundwater and in direct runoff from the ground. Determination of the quantity of water is not within the scope of this course and involves an entire discipline. However, there are a couple of points I do want to make about surface waters, which we will not discuss in relation to discharges. The first is that the surface waters can contain anything from suspended solids to bacteria, from nutrients to logs and automobile bodies.

The second is that you have to look closely at the surface waters when you are planning a discharge. Chances are good that you will be discharging to a surface supply and, perhaps, someone else's drinking water.

Permits for facilities discharging to the surface waters are often written on a "net" discharge basis, or on an average basis, and that represents a potential danger to the plant. Permitting engineers only know one type of distribution: regular or normal distribution. *Hydrologic events such as rainfall and runoff are not normally distributed. This is also true for contaminant loads.* There is ample evidence that the hydrologic events, which generate river flow and river water quality can be modeled by either a log-normal distribution or a Weibull type III distribution. We will talk about some of those distributions later. However, the point is that the permits are written around average statistics, which do not apply, and if you are not careful that can get you into a lot of trouble.

Storm Water

Storm water consists of rainfall, snow melt, hail, and other types of precipitation. It washes the atmosphere and transfers air contaminants into the rain. Hence, stormwater often contains carbonates and sulfates if the air is in an industrial area and the air pollution is bad. Acid rain is really harmful, and it can affect your plant operation.

In Louisiana a few years ago, a company, the author was working for, attempted to get the regulatory community to issue a permit allowance for acid rainfall in the plant discharge permit. The company had documented that the pH decreased substantially whenever they had a rainfall. This is the same type of documentation that the United States has experienced in the Acid Rain Debate where coal-fired boilers in the Midwest are emitting enough sulfur dioxide—which converts to sulfurous and then sulfuric acid—to change the pH of the rainfall and cause lakes to change their pH, and as a result, their ecosystems. At that time, the request was reasonable, but it was rejected.

In a plant environment, one has to consider the design of the sewer system and the response time for sewer events. In general, if you are monitoring your plant sewer system, you will find a dramatic increase in pollution conveyed to the treatment plant because of the "first flush" phenomenon. As the storm continues, you may find unusual discharges because someone in the plant has decided to "get rid of" that tank of chemical X during the storm, hoping that no one will notice.

Loading operations incidents are also potentially troublesome at this time as well. This is both because the equipment may have physical problems during the rain, and because the operator really does not want to go out into the rain or stay in the rain to monitor the equipment the same way he will in dry weather. Wherever possible, it might be advisable to have loading areas covered so that you will not have to treat the volume of the spill plus the volume of the storm water if there is a spill somewhere close to the time of a rainfall event.

WATER QUALITY

Water is often ranked by its quality. However, there are many different measures of water quality, and the quality of the water often depends upon its use. Water used for drinking tastes flat if it does not have some small quantity of minerals and dissolved oxygen in it. However, that same water so preferred for drinking is terrible for use in a boiler. Similarly, moderate quantities of sulfate in drinking water will cause osmotic diarrhea in sensitive individuals as well as boiler corrosion.[3] Dissolved oxygen corrodes boiler tubes, and calcium salts will form deposits on the tubes, reducing the heat transfer efficiency.

Potable water generally has total sodium salt concentrations below 200 mg/l. Salt concentrations greater than 70 mg/l cause the water to taste salty, and above 5000 mg/l of sodium, water is considered brackish and can cause problems with osmotic pressure in human beings.[4] When the sodium concentration is above 100 there is some small risk to human beings sensitive to sodium in their diet, and various regulatory agencies have suggested maximum sodium concentrations between 100 and 160 mg/l for drinking water.

[3]EPA suggests that diarrhea can be caused by sulfate levels of more than 650 mg/l in infants and more than 1400 mg/l in adults. For more details see http://www.epa.gov/safewater/standard/sfstudy.pdf.
[4]The actual definition of brackish water is between 0.5% and about 1.8% salt (500–18,000 parts per thousand).

Water quality, especially freshwater quality, is often classified by its uses: recreational, drinking, fishing, and recharge. It is important to understand how the water upstream and downstream is being used because the downstream use will often dictate the overall water quality – and that will affect the discharge criteria for water discharge.

Example: There is an old joke about the quickest way to eliminate water pollution: Have the Municipalities build their drinking water intakes downstream of the effluent from the wastewater treatment plant. The most ironic thing about the joke is that with the current water shortages, the need for recycling is growing to the point where highly treated effluent are being put back into the drinking water reservoir in several communities. This effluent has a better quality than that of the reservoir, and the joke is really becoming true.

According to the current water quality control schemes in use in the United States, the highest use for water is for human consumption. Water for human consumption must meet two sets of standards: the river or stream source standards and the Primary and Secondary Water Quality Standards published by the USEPA and by the various States.

WATER QUALITY REGULATIONS—LEGAL STRUCTURE

Water quality standards are dependent upon the purposes for which the water is used.

Example: Cooling water's principal characteristic must be temperature and to a lesser extent lack of corrosiveness.

The United States, the United Nations and most of the countries have water quality standards for drinking water. Many countries have water quality standards for fishing and swimming waters depending upon their uses.

When dealing with the United States, it is important to remember that the goal of the USEPA is to have fishable and swimmable waters (fishing and recreation uses) for all U.S. waters. That goal has been in place since 1972 and still has not been achieved.

It is important to note that surface water quality standards are widely different from place to place and depend upon the use of the water.

The following pages contain the excerpts from the State of Georgia Water Quality Standards for Surface Waters. They have been annotated for better understanding because it is important to understand how the regulations are structured so that you have an adequate basis for knowing what the regulatory community is charged to do, and what their priorities are. A few minutes study will give you an idea of how the philosophy of water quality goals and effluent limitations has developed.

RULES AND REGULATIONS FOR WATER QUALITY CONTROL

Chapter 391-3-6 Revised—July 2000

Several things have been annotated to provide you with a flavor for the subject and its complexity. Understand that the text was written by lawyers, and normally it is interpreted by the rest of us (unimportant preamble has been deleted).

(4) Water Use Classifications. Water use classifications for which the criteria of this Paragraph are applicable are as follows:
   (a) Drinking Water Supplies
   (b) Recreation
   (c) Fishing, Propagation of Fish, Shellfish, Game and Other Aquatic Life
   (d) Wild River
   (e) Scenic River
   (f) Coastal Fishing
(5) General Criteria for All Waters. The following criteria are deemed to be necessary and applicable to all waters of the State:

Note that the order of the text has been arranged in the order of priority.

General Provisions and Catchall Regulations

(a) All waters shall be free from materials associated with municipal or domestic sewage, industrial waste or any other waste which will settle to form sludge deposits that become putrescent, unsightly or otherwise objectionable.
(b) All waters shall be free from oil, scum and floating debris associated with municipal or domestic sewage, industrial waste or other discharges in amounts sufficient to be unsightly or to interfere with legitimate water uses.
(c) All waters shall be free from material related to municipal, industrial or other discharges which produce turbidity, color, odor or other objectionable conditions which interfere with legitimate water uses.
(d) **Turbidity**. The following standard is in addition to the narrative turbidity standard in Paragraph 391-3-6-.03(5)(c) above:
All waters shall be free from turbidity which results in a substantial visual contrast in a water body due to a man-made activity. The upstream appearance

of a body of water shall be as observed at a point immediately upstream of a turbidity-causing man-made activity. That upstream appearance shall be compared to a point which is located sufficiently downstream from the activity so as to provide an appropriate mixing zone. For land disturbing activities, proper design, installation, and maintenance of best management practices and compliance with issued permits shall constitute compliance with Paragraph 391-3-6-.03(5)(d).

(e) All waters shall be free from toxic, corrosive, acidic and caustic substances discharged from municipalities, industries or other sources, such as nonpoint sources, in amounts, concentrations or combinations which are harmful to humans, animals or aquatic life.

In the above several things are important: the use of catchall provisions, see underlined material in Sections b, c, and d, and the use of a mixing zone in paragraph d.

The mixing zone is a very interesting concept because it is an artificial zone where dilution of the effluent is specifically permitted. Generally mixing zones are up to one-third of the volume of the stream and provide an undetermined length. The regulatory purpose of a mixing zone is to insure that the stream quality standards are not applied to the effluent at the point of discharge. However, when there is a small stream, which is intermittent, the tighter effluent standards will apply to the effluent.

Specific Chemical Limitations (Broadly Applicable)

(I) Instream concentrations of the following chemical constituents which are considered to be other toxic pollutants of concern in the State of Georgia shall not exceed the criteria indicated below under 7-day, 10-year minimum flow (7Q10) or higher stream flow conditions except within established mixing zones:

The term "7Q10" has an extremely conservative definition — it is the lowest flow that occurs for 7 consecutive days once in every 10 years. By definition this is during a drought cycle (summer) when water temperatures are the highest and dissolved oxygen is the lowest. This is also the baseline for all water quality standards in freshwater streams and rivers.

Specific Chemicals Followed by Concentrations

1. 2,4-Dichlorophenoxyacetic acid (2,4-D) 70 µg/l
2. Methoxychlor 0.03 µg/l*
3. 2,4,5-Trichlorophenoxy propionic acid (TP Silvex) 50 µg/l

(ii) Instream concentrations of the following chemical constituents listed by the U.S. Environmental Protection Agency as toxic priority pollutants

pursuant to Section 307(a)(1) of the Federal Clean Water Act (as amended) shall not exceed the acute criteria indicated below under 1-day, 10-year minimum flow (1Q10) or higher stream flow conditions and shall not exceed the chronic criteria indicated below under 7-day, 10-year minimum flow (7Q10) or higher stream flow conditions except within established mixing zones or in accordance with site specific effluent limitations developed in accordance with procedures presented in 391-3-6-.06. Unless otherwise specified, the criteria below are listed in their total recoverable form. Because most of the numeric criteria for the metals below are listed as the dissolved form, total recoverable concentrations of metals that are measured instream will need to be translated to the dissolved form in order to compare the instream data with the numeric criteria. This translation will be performed using guidance found in "Guidance Document of Dynamic Modeling and Translators August 1993" found in Appendix J of EPA's
Water Quality Standards Handbook: Second Edition, EPA-823-B-94-005a or by using other appropriate guidance from EPA.

Acute Chronic

1. Arsenic
   (a) Freshwater 50 µg/l 50 µg/l 1 1
   (b) Coastal and Marine Estuarine Waters 69 µg/l 36 µg/l 1 1
2. Cadmium
   (a) Freshwater 1.7 µg/l 0.62 µg/l 1,2,3 1,2,3
   (b) Coastal and Marine Estuarine Waters 43 µg/l 9.2 µg/l 1 1,2
3. Chromium III
   (a) Freshwater 310 µg/l 100 µg/l 1,3 1,3
   (b) Coastal and Marine Estuarine Waters – –
4. Chromium VI
   (a) Freshwater 16 µg/l 11 µg/l 1 1
   (b) Coastal and Marine Estuarine Waters 1,100 µg/l 50 µg/l 1 1
5. Copper
   (a) Freshwater 8.8 µg/l 6.2 µg/l 1,2,3 1,2,3
   (b) Coastal and Marine Estuarine Waters 2.4 µg/l 2.4 µg/l 1,2 1,2
6. Lead
   (a) Freshwater 30 µg/l 1.2 µg/l 1,3 1,2,3
   (b) Coastal and Marine Estuarine Waters 130 µg/l 5.3 µg/l 1 1,2
7. Mercury
   (a) Freshwater 0.012 µg/l – 2
   (b) Coastal and Marine Estuarine Waters 0.025 µg/l – 2
8. Nickel
   (a) Freshwater 790 µg/l 88 µg/l 1,3 1,3
   (b) Coastal and Marine Estuarine Waters 74 µg/l 8.2 µg/l 1 1,2

9. Selenium
   (a) Freshwater – 5.0 µg/l 2
   (b) Coastal and Marine Estuarine Waters – 71 µg/l 1
10. Silver – 4 4
11. Zinc
   (a) Freshwater 64 µg/l 58 µg/l 1,3 1,3
   (b) Coastal and Marine Estuarine Waters 90 µg/l 81 µg/l 1 1

The in-stream criterion is expressed in terms of the dissolved fraction in the water column. Conversion factors used to calculate dissolved criteria are found in 40 CFR 131.36 and the Federal Register, Volume 60, No. 86, Thursday, May 4, 1995. The in-stream criterion is lower than the EPD laboratory detection limits. The aquatic life criteria for these metals are expressed as a function of total hardness (mg/l) in a water body. Values in the table above assume a hardness of 50 mg/l $CaCO_3$. For other hardness values, the following equations from 40 CFR 131.36 should be used. The minimum hardness allowed for use in these equations shall not be less than 25 mg/l, as calcium carbonate and the maximum shall not be greater than 400 mg/l as calcium carbonate.

Cadmium
acute criteria = (e)(1.136672−[(ln hardness)(0.041838)] µg/l (1.128[ln(hard-ness)] − 3.828)
chronic criteria = (e)(1.101672-[(ln hardness)(0.041838)] µg/l (0.7852[ln (hardness)] − 3.490)

Chromium III
acute criteria = (e) (0.316) µg/l (0.8190[ln(hardness)] + 3.688)
chronic criteria = (e)(0.860) µg/l (0.8190[ln(hardness)] + 1.561)
Copper
acute criteria = (e)(0.96) µg/l (0.9422[ln(hardness)] − 1.464)
chronic criteria = (e)(0.96) µg/l (0.8545[ln(hardness)] − 1.465)

Lead
acute criteria = (e)(1.46203 − [(ln hardness)(0.145712)]) µg/l (1.273[ln (hardness) − 1.460)
chronic criteria = (e)(1.46203 − [(ln hardness)(0.145712)]) µg/l (1.273[ln (hardness) − 4.705)

Nickel
acute criteria = (e)(.998) µg/l (0.8460[ln(hardness)] + 3.3612)
chronic criteria = (e)(.997) µg/l (0.8460[ln(hardness)] + 1.1645)

Zinc
acute criteria = (e)(0.978) µg/l (0.8473[ln(hardness)] + 0.8604)
chronic criteria = (e)(0.986) µg/l (0.8473[ln(hardness)] + 0.7614)
This pollutant is addressed in 391-3-6-.06.4

Specific Organic Chemicals or Priority Pollutants (Established by Federal Decree)

(iii) Instream concentrations of the following chemical constituents listed by the U.S. Environmental Protection Agency as toxic priority pollutants pursuant to Section 307(a)(1) of the Federal Clean Water Act (as amended) shall not exceed criteria indicated below under 7-day, 10-year minimum flow (7Q10) or higher stream flow conditions except within established mixing zones or in accordance with site specific effluent limitations developed in accordance with procedures presented in 391-3-6-.06.

Specific Priority Pollutants (*Followed by Compounds—These Are Embodied in U.S. Federal Law*)

1. Chlordane
    (a) Freshwater 0.0043 μg/l*
    (b) Coastal and Marine Estuarine Waters 0.004 μg/l*

2. Cyanide
    (a) Freshwater 5.2 μg/l*
    (b) Coastal and Marine Estuarine Waters 1.0 μg/l*

3. Dieldrin 0.0019 μg/l*

4. 4,4′-DDT 0.001 μg/l*

5. a-Endosulfan
    (a) Freshwater 0.056 μg/l*
    (b) Coastal and Marine Estuarine Waters 0.0087 μg/l*

6. b-Endosulfan
    (a) Freshwater 0.056 μg/l*
    (b) Coastal and Marine Estuarine Waters 0.0087 μg/l*

7. Endrin 0.002 μg/l*

8. Heptachlor
    (a) Freshwater 0.0038 μg/l*
    (b) Coastal and Marine Estuarine Waters 0.0036 μg/l*

9. Heptachlor Epoxide
    (a) Freshwater 0.0038 μg/l*
    (b) Coastal and Marine Estuarine Waters 0.0036 μg/l*

10. Lindane [Hexachlorocyclohexane (g-BHC-Gamma)] 0.08 μg/l

11. Pentachlorophenol
    (a) Freshwater 2.1 μg/l*
    (b) Coastal and Marine Estuarine Waters 7.9 μg/l*

12. PCB-1016 0.014 μg/l

13. PCB-1221 0.014 μg/l

14. PCB-1232 0.014 μg/l

15. PCB-1242 0.014 μg/l

16. PCB-1248 0.014 µg/l
17. PCB-1254 0.014 µg/l
18. PCB-1260 0.014 µg/l
19. Phenol 300 µg/l
20. Toxaphene 0.0002 µg/l*

*The in-stream criterion is lower than the EPD[5] laboratory detection limits.

(iv) Instream concentrations of the following chemical constituents listed by the U. S. Environmental Protection Agency as toxic priority pollutants pursuant to Section 307(a)(1) of the Federal Clean Water Act (as amended) shall not exceed criteria indicated below under annual average or higher stream flow conditions:

1. Acenaphthene **
2. Acenaphthylene **
3. Acrolein 780 µg/l
4. Acrylonitrile 0.665 µg/l
5. Aldrin 0.000136 µg/l
6. Anthracene 110000 µg/l
7. Antimony 4308 µg/l
8. Arsenic 50 µg/l
9. Benzidine 0.000535 µg/l
10. Benzo(a)Anthracene 0.0311 µg/l
11. Benzo(a)Pyrene 0.0311 µg/l
12. 3,4-Benzofluoranthene 0.0311 µg/l
13. Benzene 71.28 µg/l
14. Benzo(ghi)Perylene **
15. Benzo(k)Fluoranthene 0.0311 µg/l
16. Beryllium **
17. a-BHC-Alpha 0.0131 µg/l
18. b-BHC-Beta 0.046 µg/l
19. Bis(2-Chloroethyl)Ether 1.42 µg/l
20. Bis(2-Chloroisopropyl)Ether 170000 µg/l
21. Bis(2-Ethylhexyl)Phthalate 5.92 µg/l
22. Bromoform (Tribromomethane) 360 µg/l
23. Carbon Tetrachloride 4.42 µg/l
24. Chlorobenzene 21000 µg/l
25. Chlorodibromomethane 34 µg/l
26. 2-Chloroethylvinyl Ether **

[5]EPD is the Environmental Protection of the State of Georgia.

27. Chlordane 0.000588 µg/l
28. Chloroform (Trichloromethane) 470.8 µg/l
29. 2-Chlorophenol **
30. Chrysene 0.0311 µg/l
31. Dibenzo(a,h)Anthracene 0.0311 µg/l
32. Dichlorobromomethane 22 µg/l
33. 1,2-Dichloroethane 98.6 µg/l
34. 1,1-Dichloroethylene 3.2 µg/l
35. 1,3-Dichloropropylene (Cis) 1700 µg/l
36. 1,3-Dichloropropylene (Trans) 1700 µg/l
37. 2,4-Dichlorophenol 790 µg/l
38. 1,2-Dichlorobenzene 17000 µg/l
39. 1,3-Dichlorobenzene 2600 µg/l
40. 1,4-Dichlorobenzene 2600 µg/l
41. 3,3'-Dichlorobenzidine 0.077 µg/l
42. 4,4'-DDT 0.00059 µg/l
43. 4,4'-DDD 0.00084 µg/l
44. 4,4'-DDE 0.00059 µg/l
45. Dieldrin 0.000144 µg/l
46. Diethyl Phthalate 120000 µg/l
47. Dimethyl Phthalate 2900000 µg/l
48. 2,4-Dimethylphenol **
49. 2,4-Dinitrophenol 14264 µg/l
50. Di-n-Butyl Phthalate 12100 µg/l
51. 2,4-Dinitrotoluene 9.1 µg/l
52. 1,2-Diphenylhydrazine 0.54 µg/l
53. Endrin Aldehyde 0.81 µg/l
54. Endosulfan Sulfate 2.0 µg/l
55. Ethylbenzene 28718 µg/l
56. Fluoranthene 370 µg/l
57. Fluorene 14000 µg/l
58. Heptachlor 0.000214 µg/l
59. Heptachlor Epoxide 0.00011 µg/l
60. Hexachlorobenzene 0.00077 µg/l
61. Hexachlorobutadiene 49.7 µg/l
62. Hexachlorocyclopentadiene 17000 µg/l
63. Hexachloroethane 8.85 µg/l
64. Indeno(1,2,3-cd)Pyrene .0311 µg/l
65. Isophorone 600 µg/l

66. Lindane [Hexachlorocyclohexane (g-BHC-Gamma)] 0.0625 µg/l
67. Methyl Bromide (Bromomethane) 4000 µg/l
68. Methyl Chloride (Chloromethane) **
69. Methylene Chloride 1600 µg/l
70. 2-Methyl-4,6-Dinitrophenol 765 µg/l
71. 3-Methyl-4-Chlorophenol **
72. Nitrobenzene 1900 µg/l
73. N-Nitrosodimethylamine 8.12 µg/l
74. N-Nitrosodi-n-Propylamine **
75. N-Nitrosodiphenylamine 16.2 µg/l
76. PCB-1016 0.00045 µg/l
77. PCB-1221 0.00045 µg/l
78. PCB-1232 0.00045 µg/l
79. PCB-1242 0.00045 µg/l
80. PCB-1248 0.00045 µg/l
81. PCB-1254 0.00045 µg/l
82. PCB-1260 0.00045 µg/l
83. Phenanthrene **
84. Phenol 4,600,000 µg/l
85. Pyrene 11,000 µg/l
86. 1,1,2,2-Tetrachloroethane 10.8 µg/l
87. Tetrachloroethylene 8.85 µg/l
88. Thallium 6.3 µg/l
89. Toluene 200000 µg/l
90. 1,2-Trans-Dichloroethylene **
91. 1,1,2-Trichloroethane 41.99 µg/l
92. Trichloroethylene 80.7 µg/l
93. 2,4,6-Trichlorophenol 6.5 µg/l
94. 1,2,4-Trichlorobenzene **
95. Vinyl Chloride 525 µg/l

** These pollutants are addressed in 391-3-6-.06.

(v) Site specific criteria for the following chemical constituents will be developed on an as-needed basis through toxic pollutant monitoring efforts at new or existing discharges that are suspected to be a source of the pollutant at levels sufficient to interfere with designated uses:
1. Asbestos
(vi) Instream concentrations of 2,3,7,8-tetrachlorodibenzo-p-dioxin (TCDD) must not exceed 0.0000012 µg/l under long-term average stream flow conditions.

(f) Applicable State and Federal requirements and regulations for the discharge of radioactive substances shall be met at all times.

(g) The dissolved oxygen criteria as specified in individual water use classifications shall be applicable at a depth of one meter below the water surface; in those instances where depth is less than two meters, the dissolved oxygen criterion shall be applied at a mid-depth. On a case specific basis, alternative depths may be specified.

(6) Specific Criteria for Classified Water Usage. In addition to the general criteria, the following criteria are deemed necessary and shall be required for the specific water usage as shown:

The following is a listing of <u>Minimum Water Quality Criteria For a Public Drinking Water Surface Supply</u>: note the differences between some of the standards above and the following.

Coliform or Bacterial Standard—the First Standard

(a) Drinking Water Supplies: Those waters approved as a source for public drinking water systems permitted or to be permitted by the Environmental Protection Division. Waters classified for drinking water supplies will also support the fishing use and any other use requiring water of a lower quality.

(I) Bacteria: For the months of May through October, when water contact recreation activities are expected to occur, fecal coliform not to exceed a geometric mean of 200 per 100 ml based on at least four samples collected from a given sampling site over a 30-day period at intervals not less than 24 hours. Should water quality and sanitary studies show fecal coliform levels from non-human sources exceed 200/100 ml (geometric mean) occasionally, then the allowable geometric mean fecal coliform shall not exceed 300 per 100 ml in lakes and reservoirs and 500 per 100 ml in free flowing freshwater streams. For the months of November through April, fecal coliform not to exceed a geometric mean of 1,000 per 100 ml based on at least four samples collected from a given sampling site over a 30-day period at intervals not less than 24 hours and not to exceed a maximum of 4,000 per 100 ml for any sample. The State does not encourage swimming in surface waters since a number of factors which are beyond the control of any State regulatory agency contribute to elevated levels of fecal coliform.

Dissolved Oxygen Standard—the Second Standard

(ii) Dissolved oxygen: A daily average of 6.0 mg/l and no less than 5.0 mg/l at all times for waters designated as trout streams by the Wildlife Resources Division. A daily average of 5.0 mg/l and no less than 4.0 mg/l at all times for water supporting warm water species of fish.

During the summer, when the water temperature is 68°F or warmer for smaller streams, the maximum dissolved oxygen concentration that the water can hold is approximately 9 mg/l. The requirement is about 67% of the maximum based upon a 30-day average.

pH Standard

(iii) pH: Within the range of 6.0–8.5.

And Catchall Physical Standards Including Temperature Increase

(iv) No material or substance in such concentration that, after treatment by the public water treatment system, exceeds the maximum contaminant level established for that substance by the Environmental Protection Division pursuant to the Georgia Rules for Safe Drinking Water.

(v) Temperature: Not to exceed 90°F. At no time is the temperature of the receiving waters to be increased more than 5°F above intake temperature except that in estuarine waters the increase will not be more than 1.5°F. In streams designated as primary trout or smallmouth bass waters by the Wildlife Resources Division, there shall be no elevation of natural stream temperatures. In streams designated as secondary trout waters, there shall be no elevation exceeding 2°F of natural stream temperatures.

Water Quality for Recreation Activities (*The Second Highest Use Category*)

(b) Recreation: General recreational activities such as water skiing, boating, and swimming, or for any other use requiring water of a lower quality, such as recreational fishing. These criteria are not to be interpreted as encouraging water contact sports in proximity to sewage or industrial waste discharges regardless of treatment requirements:

    (I)   Bacteria: Fecal coliform not to exceed the following geometric means based on at least four samples collected from a given sampling site over a 30-day period at intervals not less than 24 hours:

    (II) (1) Coastal waters 100 per 100 ml

          (2) All other recreational waters 200 per 100 ml

          (3) Should water quality and sanitary studies show natural fecal coliform levels exceed 200/100 ml (geometric mean) occasionally in high quality recreational waters, then the allowable geometric mean fecal coliform level shall not exceed 300 per 100 ml in lakes and reservoirs and 500 per 100 ml in free flowing fresh water streams.

(ii) Dissolved Oxygen: A daily average of 6.0 mg/l and no less than 5.0 mg/l at all times for waters designated as trout streams by the Wildlife Resources

Division. A daily average of 5.0 mg/l and no less than 4.0 mg/l at all times for waters supporting warm water species of fish.

(iii) pH: Within the range of 6.0–8.5.

(iv) Temperature: Not to exceed 90°F. At no time is the temperature of the receiving waters to be increased more than 5°F above intake temperature except that in estuarine waters the increase will not be more than 1.5°F. Instreams designated as primary trout or smallmouth bass waters by the Wildlife Resources Division, there shall be no elevation of natural stream temperatures. Instreams designated as secondary trout waters, there shall be no elevation exceeding 2°F natural stream temperatures.

Water Quality Catchall For Fishing (*Third Priority Use*)

(c) Fishing: Propagation of Fish, Shellfish, Game and Other Aquatic Life; secondary contact recreation in and on the water; or for any other use requiring water of a lower quality:

(i) Dissolved Oxygen: A daily average of 6.0 mg/l and no less than 5.0 mg/l at all times for water designated as trout streams by the Wildlife Resources Division. A daily average of 5.0 mg/l and no less than 4.0 mg/l at all times for waters supporting warm water species of fish.

(ii) pH: Within the range of 6.0–8.5.

(iii) Bacteria: For the months of May through October, when water contact recreation activities are expected to occur, fecal coliform not to exceed a geometric mean of 200 per 100 ml based on at least four samples collected from a given sampling site over a 30-day period at intervals not less than 24 hours. Should water quality and sanitary studies show fecal coliform levels from non-human sources exceed 200/100 ml (geometric mean) occasionally, then the allowable geometric mean fecal coliform shall not exceed 300 per 100 ml in lakes and reservoirs and 500 per 100 ml in free flowing freshwater streams. For the months of November through April, fecal coliform not to exceed a geometric mean of 1,000 per 100 ml based on at least four samples collected from a given sampling site over a 30-day period at intervals not less than 24 hours and not to exceed a maximum of 4,000 per 100 ml for any sample. The State does not encourage swimming in surface waters since a number of factors which are beyond the control of any State regulatory agency contribute to elevated levels of fecal coliform. For waters designated as approved shellfish harvesting waters by the appropriate State agencies, the requirements will be consistent with those established by the State and Federal agencies responsible for the National Shellfish Sanitation Program. The requirements are found in the National Shellfish Sanitation Program Manual of Operation, Revised 1988, Interstate Shellfish Sanitation Conference, U. S. Department of Health and Human Services (PHS/FDA), and the Center for Food Safety and Applied Nutrition. Streams designated as generally supporting shellfish are listed in Paragraph 391-3-6-.03(14).

(iv) Temperature: Not to exceed 90°F. At no time is the temperature of the receiving waters to be increased more than 5°F above intake temperature except that in estuarine waters the increase will not be more than 1.5°F. In streams designated as primary trout or smallmouth bass waters by the Wildlife Resources Division, there shall be no elevation of natural stream temperatures. In streams designated as secondary trout waters, there shall be no elevation exceeding 2°F natural stream temperatures.

Other Uses

(d) Wild River: For all waters designated in 391-3-6-.03(13) as "Wild River," there shall be no alteration of natural water quality from any source.

(e) Scenic River: For all waters designated in 391-3-6-.03(13) as "Scenic River," there shall be no alteration of natural water quality from any source.

(f) Coastal Fishing: This classification will be applicable to specific sites when so designated by the Environmental Protection Division.

For waters designated as "Coastal Fishing," site specific criteria for dissolved oxygen will be assigned and detailed by footnote in Section 391-3-6-.03(3), "Specific Water Use Classifications." All other criteria and uses for the fishing use classification will apply for coastal fishing.

Exemptions and Other General Requirements

(7) Natural Water Quality. It is recognized that certain natural waters of the State may have a quality that will not be within the generator specific requirements contained herein. This is especially the case for the criteria for dissolved oxygen, temperature, pH and fecal coliform. NPDES permits and best management practices will be the primary mechanisms for ensuring that discharges will not create a harmful situation.

(8) Treatment Requirements. Notwithstanding the above criteria, the requirements of the State relating to secondary or equivalent treatment of all waste shall prevail. The adoption of these criteria shall in no way preempt the treatment requirements.

(9) Streamflows. Specific criteria or standards set for the various parameters apply to all flows on regulated streams. On unregulated streams, they shall apply to all streamflows equal to or exceeding the 7-day, 10-year minimum flow (7Q10) and/or the 1-day, 10-year minimum flow. All references to 7-day, 10-year minimum flow (7Q10) and 1-day, 10-year minimum flow also apply to all flows on regulated streams. All references to annual average stream flow also apply to long-term average stream flow conditions.

(10) Mixing Zone. Effluents released to streams or impounded waters shall be fully and homogeneously dispersed and mixed insofar as practical with the main flow or water body by appropriate methods at the discharge point. Use of a reasonable and limited mixing zone may be permitted on receipt of satisfactory evidence that such a zone is necessary and that it will not create an

objectionable or damaging pollution condition. Protection from acute toxicity shall be provided within any EPD designated mixing zone to ensure a zone of safe passage for aquatic organisms. The procedure is as described in paragraph 391-3-6-.06(4)(d)(5)(vi), except that the numerical pass/fail criteria applies to the end-of-pipe without the benefit of dilution provided by the receiving stream.

(11) Toxic Pollutant Monitoring. The Division will monitor waters of the State for the presence or impact of Section 307(a)(l) Federal Clean Water Act toxic pollutants, and other priority pollutants. The monitoring shall consist of the collection and assessment of chemical and/or biological data as appropriate from the water column, from streambed sediments, and/or from fish tissue. Specific stream segments and chemical constituents for monitoring shall be determined by the Director on the basis of the potential for water quality impacts from toxic pollutants from point or nonpoint waste sources. Singularly or in combination, these constituents may cause an adverse effect on fish propagation at levels lower than the criteria. Instream concentrations will be as described in 391-3-6-.03 (5)(e). Additional toxic substances and priority pollutants will be monitored on a case specific basis using Section 304(a) Federal Clean Water Act guidelines or other scientifically appropriate documents.

(12) Fecal Coliform Criteria. The criteria for fecal coliform bacteria provide the Regulatory framework to support the USEPA requirement that States protect all waters for the use of primary contact recreation or swimming. This is a worthy national goal, although potentially unrealistic with the current indicator organism, fecal coliform bacteria, in use today. To assure that waters are safe for swimming indicates a need to test waters for pathogenic bacteria. However, analyses for pathogenic bacteria are expensive and results are generally difficult to reproduce quantitatively. Also, to ensure the water is safe for swimming would require a whole suite of tests be done for organisms such as Salmonella, Shigella, Vibrio, etc. as the presence/absence of one organism would not document the presence/absence of another. This type of testing program is not possible due to resource constraints. The environmental community in the United States has based the assessment of the bacteriological quality of water on testing for pathogenic indicator organisms, principally the coliform group. The assessment of streams, rivers, lakes, and estuaries in Georgia and other States is based on fecal coliform organisms. Coliform bacteria live in the intestinal tract of warm blooded animals including man. These organisms are excreted in extremely high numbers, averaging about 1.5 billion coliform per ounce of human feces. Pathogenic bacteria also originate in the fecal material of diseased persons. Therefore, waters with high levels of fecal coliform bacteria represent potential problem areas for swimming. However, there is no positive scientific evidence correlating elevated fecal coliform counts with transmission of enteric diseases. In addition, these bacteria can originate from any warm blooded animal or from the soil. Monitoring programs have documented fecal coliform levels in excess of the criteria in

many streams and rivers in urban areas, agricultural areas, and even in areas not extensively impacted by man such as national forest areas. This is not a unique situation to Georgia as similar levels of fecal coliform bacteria have been documented in streams across the nation. The problem appears to lie in the lack of an organism which specifically indicates the presence of human waste materials which can be correlated to human illness. Other organisms such as the Enterococci group and E. coli have been suggested by the USEPA as indicator organisms. However, testing using these organisms by States and the USEPA has indicated similar problems with these indicator organisms.

The Environmental Protection Division will conduct a monitoring project from 1993 through 1995 to evaluate the use of E. coli and Enterococci as indicators of bacteriological quality in Georgia. The Environmental Protection Division will also conduct studies to determine if a better human specific indicator can be found to replace current indicator organisms.

(13) Specific Water Use Classifications. Beneficial water uses assigned by the State to all surface waters. These classifications are scientifically determined to be the best utilization of the surface water from an environmental and economic standpoint. Streams and stream reaches not specifically listed are classified as Fishing.

The specific classifications are as follows:

Specific Stream Classification Standards

Savannah River Basin Classification
Chattooga River Georgia—North Carolina State Line to Tugaloo Reservoir - Wild and Scenic
West Fork Chattooga Confluence of Overflow Creek and Clear Creek to confluence with Chattooga River (7.3 mi.)—Wild and Scenic

Of course, there are more detailed water quality standards, but those are for georgia and are not really of interest anywhere else.

APPLICATIONS

The single "Most Important Water Quality Parameter" is probably dissolved oxygen. Before we go on to drinking water, we will look at some of the concepts behind this because it impacts what and how we treat our wastes.

Dissolved oxygen is important because it determines what happens in the water, whether the water is "clean or dirty," and dictates our perception of water quality.

Dissolved oxygen is measured by one of the several techniques. The preferred method is by Oxygen Electrode, but the older Winkler test is often used to calibrate the electrodes. The Winkler test uses a divalent manganese solution followed by a strong alkali to develop manganese hydroxide. Iodine is then added along with starch, and the mixture is then back titrated with a standard solution of sodium thiosulfate. The end point is very sharp and the accuracy of the colorimetric test is accurate to about 0.01 mg/l.

BOD_5 is the 5-day Biochemical Oxygen Demand. It is a measure of how much dissolved oxygen is consumed by an acclimatized waste stream in 5 days by the organic carbon material in the waste stream. It is the broad measure of the strength of the organic matter in a waste stream. The test is conducted by preparation of a known quantity of nutrient dilution water rich in dissolved oxygen. Known aliquots of waste are measured and placed into special bottles where the seal prevents air from diffusing oxygen that is diffusing into the sample. The sample is then incubated at 20°C for 5 days in the dark. At the end of 5 days, the dissolved oxygen drop is measured and the oxygen demand of the waste is calculated from the size of the aliquot of waste entered into the bottle. According to legend, the BOD test was developed in England where no river required more than 5 days to flow to the sea. In the United States, domestic sewage has a BOD_5 of between 100 mg/l and 160 mg/l. The test is used as a reporting parameter, but it is useless for control and process purposes. Few wastewater treatment plants have a retention capacity of 5 days, and the majority of the plants are at a retention capacity of 12 h or less. The information provided by the BOD test is primarily for historical information because by the time the results are known, the waste from the effluent would be from at least 5 days travel time downstream. However, in the United States there is a dogged reluctance to utilize anything but BOD for measurement and reporting purposes, despite the fact that it is useless as a control parameter.

The accuracy of the BOD test is also questionable. *Standard Methods* reports the accuracy of the test as about ±30.5 mg/l at a waste strength of 198 mg/l. The reported minimum accuracy of the test is 2 mg/l, but in practical terms, numbers below 10 mg/l are all in the same range of unreliability.

The BOD test must also be corrected for nitrification and waste acclimatization. Ammonia will oxidize and form nitrates, consuming oxygen in the process. This will cause the BOD to be overstated. The correction for this is an addition of ammonium chloride to the test bottle. The ammonium chloride will prevent nitrification. The challenge of waste acclimatization is more difficult. According to the test method, the dilution and seed water must contain quantities of bacteria that have grown on or have been acclimatized to the wastes being tested. For normal sewage this is not a problem. Industrial wastes often have specialized chemicals for which the bacterial population has not developed enzymes required for hydrolysis of the waste. In an unacclimatized waste, or one that contains traces of bactericides, the BOD test will report low values.

COD is chemical oxygen demand and is measured chemically by digestion with acid. There are two types of COD in use and one must be careful of the method. The United States uses a potassium dichromate digestion with a mercury catalyst. The COD test overstates the oxygen consumption by about 20%–50% on the basis of BOD measurements. Depending upon the waste stream, there may be a consistent relationship between COD and BOD, but it is highly waste dependent.

Be careful in comparing COD results from different countries. Germany and most of the Europe use a sulfuric acid digestion, which gives substantially different results and may be even stronger than the dichromate test method.

The COD test is determined in about 3 h from start to finish. As such it is a useful control parameter for oxidation operations, and given a consistent waste stream, a very good relationship can be developed between COD and BOD. The COD can also be used as a predictor of the BOD.

Sample calculation: Refer to the table on dissolved oxygen for an example of the maximum concentration of oxygen at any temperature. The rule of thumb is that the dissolved oxygen at saturation and sea level at about 20°C is about 9 mg/l. It is possible to supersaturate the oxygen in water, but it is rare without substantial turbulence.

In many rivers it is necessary to maintain a minimum dissolved oxygen concentration of about 2 mg/l. If the water gets below 2 mg/l, the fish die, and if below 0 ppm, foul smells and benthic organisms develop. An anaerobic stream is not pretty. As the dissolved oxygen level goes to or below zero, the nitrate is reduced to nitrite and then to ammonia and gaseous nitrogen, and the sulfates are reduced to H_2S. In time a stream may recover, but it is a slow and difficult process.

SAMPLE PROBLEM

Given that a stream may have 7 ppm dissolved oxygen (DO) and be flowing at 100 Cu M/h, how much waste can we place into the stream before it goes below 3 ppm? Our waste stream has 250 ppm BOD.

SOLUTION

Given that the flow is 100 CuM/h and the minimum DO is 3 ppm, which means that we have 4 ppm that we can use, if the regulatory authority will allow us to use the full 4 ppm, and if our waste is stable and does not vary above the 100 mg/l oxygen demand value.

So, 4 mg/l = 4 ppm. Since a cubic meter contains 1000 l, it means that the oxygen load the stream carries, which is available to us, is: $4 \times 100 \times 1000$ mg $= 400,000$ mg/h $= 400$ g/h.

Our waste stream has a strength of 250 mg/l. So by comparison, we can discharge 400,000 mg/h/250 mg/l = 1600 l/h or about 7.5 gallons/min. If the low flow in the river at 7Q10 is only 20 CuM/H then the regulatory authority will only allow us to discharge about 320 l/h or about 1.41 gallons/min.

However, if the regulatory authority only allows us to use one-third the capacity of the stream and hold the other two-thirds in reserve, then the calculations would look like the following:

Minimum concentration of oxygen required	= 4 mg/l
7Q10 flow	= 20 CuM/H
Available oxygen at 7Q10 (7 mg/l – 4 mg/l)	= 3 mg/l
Available oxygen mg/l/h ($3 \times 20 \times 1000$)	= 60,000
Waste allocation = Available oxygen/reserve factor	= 20,000
Allowable waste discharge = 20,000/250	= 80 l/h
Flow rate 80/3.785/60	= 0.352 gallons/min

DRINKING WATER QUALITY STANDARDS: USA AND INTERNATIONAL STANDARDS

The USEPA, the European Union (EU) and the World Health Organization (WHO) all have different sets of drinking water standards. The difference between a standard and a goal or a criterion is that neither of the latter two is enforceable but is merely an objective. With the rise of the Organic Chemical Industry in the past century, and the increased detection abilities of analytical equipment, we can today find compounds in drinking water that were not detectable even 10 years ago.

The purpose of setting drinking water standards is for the protection of public health. General criteria for setting the standards are based on the protection of the most sensitive segment of the population and the prevention of "additional" diseases specific to the population. This concept of "additional" ailment leads to the development of statistical arguments and analyses in the process of goal setting. The most common measurement used is "excess cancers per N people." The N is most often between 10,000 and 1,000,000. The concept of excess cancers is, at the least, controversial because it assumes that one can detect the differences between normal cancer rates and excess cancer rates based upon animal studies and modeling. Many of the current water quality limits were developed using the "One Hit Model" where a laboratory animal, quite often a mouse, is exposed to certain chemicals, and the exposure rates and cancer rates are measured and equated to human exposure and cancer rates. Additional factors are often used in setting the standards as well.

The USEPA has been known to add conservative exposure criteria when formulating the standard, including a resident population who takes their drinking water only from one source, who feeds on fish from that source, who showers (for volatile chemicals) daily using that water, and cooks using that water, for a period of 70 years. These assumptions can be and have been questioned unsuccessfully, as they are extremely conservative and have the effect of reducing exposure levels well beyond the measurable values.

Recently (2001), the cost benefits of drinking water standards have begun to be re-examined. A recent move by the EPA to reduce the drinking water concentrations or maximum contaminant levels (MCLs) for arsenic to 5 ug/l was rejected as being too expensive. The rationale posed by EPA was that the proposed arsenic standard would cost between $28 and $85 per user household (EPA Cost Data) after legal challenges and political review because the cost of protecting one individual life from arsenic exposure (the benefit) was determined to be excessive, and implementing the standard would have cost between $700,000,000 and $1,460,000,000 per year as compared with the EPA's cost estimate of approximately $389,000,000 per year. The EPA cost estimates were about one-third or less than the corresponding estimates prepared by the American Water Works Association (AWWA). According to the comments supplied by the City of Albuquerque, NM, the cost per life saved was estimated at $4.7 billion per year, approximately 770 times higher than the EPA's current regulatory cost benefit of $6.1 million per year per life saved.[6]

From the comments submitted, the EPA's proposed regulatory scheme was also technologically flawed, in that the technology proposed by EPA for attainment of the arsenic limits was also deemed to be technologically unfeasible. The AWWA and others prevailed because they had better cost data, and because the EPA had prepared the recommended standards without adequate internal and external review of the technology and the costs. The burden on individual water treatment facilities for monitoring and treatment was also considered excessive. The proposed EPA standard was reviewed and reissued at a 10 µg/l arsenic concentration and was deemed both attainable and economically affordable, if not without some grumbling from the water utilities.

---

[6]For a discussion of the issues, see "Comments on EPA's Proposed Arsenic Rule..." submitted by City of Albuquerque, NM, September 20, 2001. Available at http://www.cabq.gov/waterresources/docs/Arsenic%20Summary.pdf.

It is possible to attain good quality water, but it has a cost.[7] Under the arsenic rules, the smaller communities, which would be the most severely impacted by the cost of providing treatment to the 10 µg/l limit are those in the western United States where arsenic occurs naturally in the groundwater at levels well above the treatment standard. When it was discovered that an aquifer in Bangladesh and in West Bengal, India contained unacceptably high levels of arsenic in the groundwater (above the 50 µg/l WHO limit) and affected over 82 million people, the proposed solution for reducing the arsenic contamination in the groundwater was substantially simpler. Because the per capita income in the affected parts of India is quite low, it was imperative to find economical methods of removing arsenic to below the WHO standard. Several novel and innovative methods were tried, and it appears that the cheapest method is to aerate the water and then run it through a bucket filled with nails or iron pellets. The arsenic is first oxidized, and then is adsorbed onto the iron. The system is inexpensive and suitable for many of the small communities in India.

Water quality varies from place to place and country to country. In an effort to establish a generally accepted level of what contaminant levels are "safe" in drinking water, regulatory agencies such as the EPA, the EU, and the Health Departments or Ministries of various countries have each established their own drinking water standards. The WHO has also established drinking water standards. Although it is difficult to establish a comparison between the differing standards, there are a number of points of commonality with regard to metals and certain organic compounds. At one time the WHO standards were substantially more lax than the United States and EU standards, but in recent years that deficiency has been corrected. A comparison of the sets of standards for WHO and USEPA is shown side by side.

There is a difference in nomenclature between the standards that require some explanation. The USEPA uses Maximum Concentration Limit (MCL) and Maximum Concentration Limit Goal (MCLG) to express the current standards. The WHO uses the language of "guideline," which is a strong suggestion but carries no legal weight because the WHO is neither a regulatory agency nor does it have a sovereign authority over any country (see Table 1.5).

---

[7]See Gurian PL, Small MJ, Lockwood JR, Schervish, M. Addressing Uncertainty and Conflicting Cost Estimates in Revising the Arsenic MCL. Environmental Science & Technology 2001, Vol. 35, pp 4414–4420.

TABLE 1.5 WHO's Guidelines for Drinking-water Quality–set up in Geneva, 1993, Which are the International Reference Point for Standard Setting and Drinking-water Safety

Element/Substance	Symbol/Formula	Normally Found in Fresh Water/Surface Water/Ground Water	Health Based Guideline by the WHO	USEPA Maximum Contaminant Level (MCL)	Maximum Contaminant Level Guideline
Aluminum	Al	<0.2 mg/l	200 µg/l		
Ammonia	NH$_4$	(up to 0.3 mg/l in anaerobic waters)	No guideline		
Antimony	Sb	<4 µg/l	5 µg/l	6 µg/l	6 µg/l
Arsenic	As	0.01 mg/l	No guideline	10 µg/l	zero
Asbestos			No guideline	7,000,000 fibers/l	7,000,000 fibers/l
Barium	Ba	<1 µg/l	300 µg/l	2000 µg/l	2000 µg/l
Berillium	Be	<1 mg/l	No guideline	4 µg/l	4 µg/l
Boron	B	<1 µg/l	300 µg/l		
Cadmium	Cd		3 µg/l	5 µg/l	5 µg/l
Chloride	Cl		250 mg/l		
Chromium	Cr^{3+}, Cr^{6+}	<2 µg/l	50 µg/l	100 µg/l total	100 µg/l
Color			No guideline		
Copper	Cu		2000 µg/l	1300 µg/l	1300 µg/l
Cyanide	CN–		70 µg/l	200 µg/l	200 µg/l
Fluoride	F	0–10 mg/l	1500 µg/l	4000 µg/l	4000 µg/l
Iron	Fe	0–50 mg/l	No guideline		
Lead	Pb		10 µg/l	< 15 µg/l	zero
Manganese	Mn		500 µg/l		
Mercury	Hg	<0.5 µg/l	1 µg/l	2 µg/l	2 µg/l
Molybdenum	Mb	<10 µg/l	70 µg/l		
Nickel	Ni	<20 µg/l	20 µg/l		

	NO₃ and NO₂					
Nitrate and nitrite				50000 µg/l total Nitrogen	Nitrate 10000 µg/l Nitrite 1000 µg/l	Nitrate 10000 µg/l Nitrite 1000 µg/l
Turbidity				No guideline		
Selenium	Se	<10 µg/l		10 µg/l	50 µg/l	50 µg/l
Silver	Ag	5–50 µg/l		No guideline		
Sodium	Na	<20000 µg/l		200000 µg/l		
Sulfate	SO₄			500000 µg/l		
Tin	Sn			No guideline		
Thallium	Th	No guideline		No guideline	2 µg/l	0.5 µg/l
Uranium	U			1400 µg/l		
Zinc	Zn			3000 µg/l		

Organic compounds

Group	Substance	Formula	WHO Health Based Guideline
Chlorinated alkanes	Carbon tetrachloride	CCl_4	2 µg/l
	Dichloromethane	CH_2CCl_2	20 µg/l
	1,1-Dichloroethane	$C_2H_4CCl_2$	No guideline
	1,2-Dichloroethane	$ClCH_2CH_2Cl$	30 µg/l
	1,1,1-Trichloroethane	CH_3CCl_3	2000 µg/l
Chlorinated ethenes	1,1-Dichloroethene	$C_2H_2Cl_2$	30 µg/l
	1,2-Dichloroethene	$C_2H_2Cl_2$	50 µg/l
	Trichloroethene	C_2HCl_3	70 µg/l
	Tetrachloroethene	C_2Cl_4	40 µg/l
Aromatic hydrocarbons	Benzene	C_6H_6	10 µg/l
	Toluene	C_7H_8	700 µg/l
	Xylenes	C_8H_{10}	500 µg/l
			300 µg/l

(Continued)

TABLE 1.5 *(Continued)*

Element/Substance	Symbol/Formula	Normally Found in Fresh Water/Surface Water/Ground Water	Health Based Guideline by the WHO	USEPA Maximum Contaminant Level (MCL)	Maximum Contaminant Level Guideline
	Ethylbenzene	C_8H_{10}		20 µg/l	
	Styrene	C_8H_8			
	Polynuclear Aromatic Hydrocarbons (PAHs)	$C_2H_3N_1O_5P_{13}$		0.7 µg/l	
Chlorinated benzenes					
	Monochlorobenzene (MCB)	C_6H_5Cl		300 µg/l	
Dichlorobenzenes (DCBs)					
	1,2-Dichlorobenzene (1,2-DCB)	$C_6H_4Cl_2$		1000 µg/l	60–75 µg/l	60–75 µg/l
	1,3-Dichlorobenzene (1,3-DCB)	$C_6H_4Cl_2$		No guideline	60–75 µg/l	60–75 µg/l
	1,4-Dichlorobenzene (1,4-DCB)	$C_6H_4Cl_2$		300 µg/l	60–75 µg/l	60–75 µg/l
	Trichlorbenzene	$C_6H_3Cl_3$		20 µg/l		
Miscellaneous organic constituents					
	Di(2-ethylhexyl)adipate (DEHA)	$C_{22}H_{42}O_4$		80 µg/l	
	Di(2-ethylhexyl)phthalate (DEHP)	$C_{24}H_{38}O_4$		8 µg/l	
	Acrylamide	C_3H_5NO		0.5 µg/l	
	Epichlorohydrin (ECH)	C_3H_5ClO		0.4 µg/l	
	Hexachlorobutadiene (HCBD)	C_4Cl_6		0.6 µg/l	
	Ethylenediaminetetr aacetic acid (EDTA)	$C_{10}H_{12}N_2O_8$		200 µg/l	
	Nitrilotriacetic acid (NTA)	$N(CH_2COOH)_3$		200 µg/l	

Organotins	Dialkyltins	R_2SnX_2	No guideline		
	Tributil oxide (TBTO)	$C_{24}H_{54}O\ Sn_2$	2 µg/l		
Pesticides	Alachlor	$C_{14}H_{20}ClNO_2$	20 µg/l	2 µg/l	zero
	Aldicarb	$C_7H_{14}N_2O_4S$	10 µg/l		
	Aldrin and dieldrin	$C_{12}H_8Cl_6$ $C_{12}H_8Cl_6O$	0.03 µg/l		
	Atrazine	$C_8H_{14}ClN_5$	2 µg/l	3 µg/l	3 µg/l
	Bentazone	$C_{10}H_{12}N_2O_3S$	30 µg/l		
	Carbofuran	$C_{12}H_{15}NO_3$	5 µg/l	40 µg/l	40 µg/l
	Chlordane	$C_{10}H_6Cl_8$	0.2 µg/l	2 µg/l	zero
	Chlorotoluron	$C_{10}H_{13}ClN_2O$	30 µg/l		
	DDT	$C_{14}H_9Cl_5$	2 µg/l		
	1,2-Dibromo-3-chloropropane	$C_3H_5Br_2Cl$	1 µg/l		
	2,4-Dichlorophenoxyacetic acid (2,4-D)	$C_8H_6Cl_2O_3$	30 µg/l		
	1,2-Dichloropropane	$C_3H_6Cl_2$	No guideline	5 µg/l	zero
	1,3-Dichloropropane	$C_3H_6Cl_2$	20 µg/l		
	1,3-Dichloropropene	$CH_3CHClCH_2Cl$	No guideline		
	Ethylene dibromide (EDB)	$BrCH_2CH_2Br$	No guideline		
	Heptachlor and heptachlor epoxide	$C_{10}H_5Cl_7$	0.03 µg/l		
	Hexachlorobenzene (HCB)	$C_{10}H_5Cl_7O$	1 µg/l		
	Isoproturon	$C_{12}H_{18}N_2O$	9 µg/l		
	Lindane	$C_6H_6Cl_6$	2 µg/l		
	MCPA	$C_9H_9ClO_3$	2 µg/l		

(*Continued*)

TABLE 1.5 (*Continued*)

Element/Substance	Symbol/Formula	Normally Found in Fresh Water/Surface Water/Ground Water	Health Based Guideline by the WHO	USEPA Maximum Contaminant Level (MCL)	Maximum Contaminant Level Guideline
	Methoxychlor	$(C_6H_4OCH_3)_2CHCCl_3$	20 µg/l		
	Metolachlor	$C_{15}H_{22}ClNO_2$	10 µg/l		
	Molinate	$C_9H_{17}NOS$	6 µg/l		
	Pendimethalin	$C_{13}H_{19}O_4N_3$	20 µg/l		
	Pentachlorophenol (PCP)	C_6HCl_5O	9 µg/l		
	Permethrin	$C_{21}H_{20}Cl_2O_3$	20 µg/l		
	Propanil	$C_9H_9Cl_2NO$	20 µg/l		
	Pyridate	$C_{19}H_{23}ClN_2O_2S$	100 µg/l		
	Simazine	$C_7H_{12}ClN_5$	2 µg/l		
	Trifluralin	$C_{13}H_{16}F_3N_3O_4$	20 µg/l		
Chlorophenoxy herbicides (excluding 2,4-D and MCPA)	2,4-DB	$C_{10}H_{10}Cl_2O_3$	90 µg/l		
	Dichlorprop	$C_9H_8Cl_2O_3$	100 µg/l		
	Fenoprop	$C_9H_7Cl_3O_3$	9 µg/l		
	MCPB	$C_{11}H_{13}ClO_3$	No guideline		
	Mecoprop	$C_{10}H_{11}ClO_3$	10 µg/l		
	2,4,5-T	$C_8H_5Cl_3O_3$	9 µg/l		
	2-Chlorophenol (2-CP)	C_6H_5ClO	No guideline		
	2,4-Dichlorophenol (2,4-DCP)	$C_6H_4Cl_2O$	No guideline		
	2,4,6-Trichlorophenol (2,4,6-TCP)	$C_6H_3Cl_3O$	200 µg/l		

2

EFFECTS OF POLLUTION

Effluent toxicity testing
Oxygen depletion—biochemical oxygen demand
Oxygen uptake in a stream—the oxygen sag equation
Biology of polluted water

There are two types of contaminant effect that you may have to deal with. The first is toxicity and the second is oxygen depletion.

Toxicity is poisoning. It occurs primarily with metals and certain types of organic chemicals. We spent some time in the first chapter discussing the subject. Toxicity can be acute or chronic. Tests for effluent quality are increasingly being defined by toxicity testing as well as by contaminant measurements in the effluent.

EFFLUENT TOXICITY TESTING

The two most common types of toxicity testing prescribed by Federal Water Quality Regulations (Code of Federal Regulations, Part 40, Section 136) are on minnows and water bugs. The tests are either static (fixed volume) or flow-through tests, which have a duration from 1 h to as long as 9 days. The flow-through tests are often longer in duration—7 days, 21 days, or 28 days. The test procedures are designed to determine any residual toxicity in the effluent, which may come from untreated chemicals, metals, or their interactions. Effluent toxicity is of special importance where the effluent

Practical Wastewater Treatment, by David L. Russell
Copyright © 2006 John Wiley & Sons, Inc.

limitation parameters in the permit do not require specific testing for the chemical compounds in the plant.

According to the Fish and Wildlife Service the purpose of Effluent Toxicity Testing is that:

(1) it produces ecologically significant results;
(2) it generates scientifically and legally defensible data;
(3) it is based on methods that are routinely available for widespread application;
(4) it is predictive;
(5) the methods are widely applicable across a range of chemicals; and
(6) the test is simple and cost-effective.

Behavioral toxicity tests, although ecologically relevant if the endpoints measured are interpretable, have met with limited success because of their intrinsic variability when replicated. The very thing that contributes to their sensitive detection capabilities can backfire if the animals are not acclimated properly or standardized test approaches are not appropriately conducted. The expenditure of time and labor required, however, can be offset by the ecologically interpretable results of such tests.... The realization that no single test approach meets all needs or answers all questions has become even more evident over the last decade. The fact is that many "tools" are needed and each should be selected and combined with others in diverse configurations depending on the contaminants of interest and the questions being addressed. Continued effort is required to further develop meaningful, cost-effective, and field-friendly methodologies to detect contaminants and their effects on aquatic biota.[1]

The most common test organisms are *daphnia magna* and *fathead minnows (Pimephales promelas)* or *sheepshead minnows*.[2] The former is a water flea and the latter a specific type of small fish.

The problems with the tests are numerous. Anyone who has an aquarium understands this well. Fleas and fish can die, sometimes for no good reason. There is also the issue about proper acclimatization of the test

[1]Henry, Mary G. http://biology.usgs.gov/s+t/SNT/noframe/co116.htm.
[2]Methods for Estimating the Chronic Toxicity of Effluents and Receiving Waters to Estuarine and Marine Organisms, Second Edition, July 1994 (EPA/600/4-91/003). This manual describes six short-term (1-h to 9-day) methods for estimating the chronic toxicity of effluents and receiving waters to five species: The sheepshead minnow, *Cyprinodon variegatus*; the inland silverside, *Menidia beryllina*; the mysid shrimp, *Mysidopsis bahia*; the sea urchin, *Arbacia punctulata*; and the red macroalga, *Champia parvula*.

organisms to the effluent. If the test organisms die from causes unrelated to exposure to the chemicals in the effluent, one may have to re-run the test to get more conclusive results, or may find oneself embroiled in a statistical argument over whether $X\%$ dilution of the effluent is toxic to aquatic life.

In a flow-through test, one of the greatest logistical problems is having enough effluent on hand to conduct the test. Most test specifications include a requirement for effluent toxicity less than a certain percentage of test organisms surviving for a period greater than the specified test period.

The typical specification in a discharge permit will look something like this:

> "The effluent toxicity shall not exceed—for an undiluted effluent on a (species) when tested for (duration) of the test."

The test conditions must measure the survival of the organisms. Variables in the test, in addition to the toxicity of the effluent, include the followings: (1) temperature, (2) pH, (3) critical and trace nutrients, (4) food supply, (5) absence of other toxic materials, and (6) adequate oxygen levels in the test tank, to name a few. The test can and often does measure synergistic effects of pollutants, and sometimes that synergy can occur with compounds already in the aquatic ecosystem. There have been a relatively small number of cases—mostly anecdotal, where the plant effluent is better than the river effluent, but causes toxicity problems when it mixes with the river because of chemicals already in the river. Unlike chemical testing, the repeatability of the tests is often open to question because of the large number of variables and the expense of conducting the test (few number of tests because of the cost).

OXYGEN DEPLETION—BIOCHEMICAL OXYGEN DEMAND

The Biochemical Oxygen Demand (BOD) test is based upon the Winkler Dissolved Oxygen Test. In it the concentration of oxygen is measured by titration of a manganous sulfate and alkaline sodium azide solution with dilute sulfuric acid in the presence of starch, which is added near the end point of the titration. The test is generally accurate to about 20 µg/l of dissolved oxygen (DO) in natural systems. Where there are a number of interferences present, modifications are available for the test. Advances in membrane and electrode technology have simplified the test procedures. Although the wet chemistry method is still the accepted reference standard, the use of DO probes has become so common that it is also accepted.

Below 20 mg/l the BOD test is considered inaccurate. The accuracy and precision of the test decrease at low BOD levels. *Standard Methods* indicates that the test is highly variable. The typical range of variability for a known standard glucose–glutamic acid solution is 198 mg/l. Inter-laboratory measurements reported in *Standard Methods* for 14 months and 421 triplicate samples indicated that the mean of the samples was 204 mg/l and it had a standard deviation of 10.4 mg/l.[3] The control limits for the sample are ±30.5 mg/l (plus or minus approximately 3 standard deviations).

Despite some regulatory trends toward issuing permits with very low BOD numbers, the statistical reliability of very low values below 20 mg/l does not exist. However, this has not stopped the regulatory community, which is regularly issuing permits with BOD_5 values less than 3 mg/l.

The test is most often run for 5 and less frequently for 20 days but under research conditions, intermediate values are also run; however, the 5-day test is the standard. The test version most often used for regulatory purposes requires that an aliquot of waste be placed in a BOD bottle and sealed to prevent air intrusion.[4] The measured oxygen depletion of the oxygen in the bottle after 5 days of incubation in a dark place at 20°C determines the BOD_5 of a waste. The Hach Company has developed a manometric test for the BOD that is, in many instances, similar to the Warburg respirometer. The Hach test uses amber glass bottles with plastic screw cap lids and magnetic stirrers in the bottles. The cap is connected to a piece of tubing, which measures the change in atmospheric pressure. Inside the bottle is a stainless steel rod that has a cup containing potassium hydroxide (KOH). The bacteria in the sample bottle feed on the waste and the waste generates CO_2 that is absorbed by the KOH. The manometer is calibrated in units equivalent to BOD_5. The practical advantage of the Hach system is that the manometer can provide a fairly rapid indication of any potential toxicity or shock-load problems, sometimes in time to allow the operator of the wastewater treatment plant to make adjustments to the system. The manometric BOD can also be used to estimate uptake coefficients and rate constants for wastewater.

The BOD test is time, temperature, nutrient, and waste acclimatization sensitive. It comprises several portions.

---

[3]APHA, AWWA & WEF. Standard Methods, 19th Edition, p. 5-3.
[4]The BOD bottle is a narrow mouth bottle with a ground glass stopper and a funnel rim around the seal. Before the bottle is put into the incubator, water is added to the funnel neck around the seal to insure that there is an air-free seal. The top of the bottle is often further covered with foil or plastic wrap to prevent evaporation from the water seal.

1. Initial demand or depletion
2. Carbonaceous demand
3. Nitrification demand

The Initial demand is measured when a dilute sample of the waste is added to the test bottle. It is the amount of oxygen depletion that occurs immediately upon sample addition. The carbonaceous demand occurs more slowly. It can be estimated by the following equation:

$$BOD_{(t)} = BOD_i(1 - e^{-kt})$$

where t is time in days, k is a determined constant, and BOD_i is the 5-day BOD of the waste. The rate coefficient k can vary anywhere from 0.2 to 0.6 but is generally about 0.2.

The BOD is a measure of the rate of biological degradation of the material. It is primarily a measure of the carbonaceous demand, but that can be misleading. The following illustrates the point of the variables and the difference in demand from nitrification. If the test is not corrected for nitrification, the waste will appear to exert a greater carbon demand than it actually does. The correction for nitrification is to add a small amount of ammonium chloride to the dilution water, in order to inhibit the nitrifiers from consuming oxygen. Otherwise, after about 5 days, most of the carbon is exhausted and nitrification begins, and the apparent BOD is higher than the actual carbonaceous demand.

Temperature plays a major role in biochemical reactions. The rate of biochemical reactions doubles for each 10°C rise in temperature up to about 30°C–40°C, at which point the bacteria are thermally inactivated, and most bacterial activity stops.

The adjustment to the rate constant is $k/k_o = e^{C_t(T-T_o)}$, where k_o is the rate coefficient at standard conditions, C_t is an adjustment coefficient, and T and T_o are measured in centigrade from a reference of 20°C.

The adjustment ratio for various temperatures and approximate values of k/k_o are shown in Table 2.1 and Figure 2.1

TABLE 2.1   Variation of k_o with Temperature

k_o	$T - T_o$					
	−15	−10	−5	0	5	10
0.3	0.0111	0.0498	0.2231	1.0000	4.4817	7.3891
0.4	0.0025	0.0183	0.1353	1.0000	7.3891	7.3891
0.5	0.0006	0.0067	0.0821	1.0000	12.1825	7.3891
0.6	0.0000	0.0009	0.0302	1.0000	33.1155	7.3891

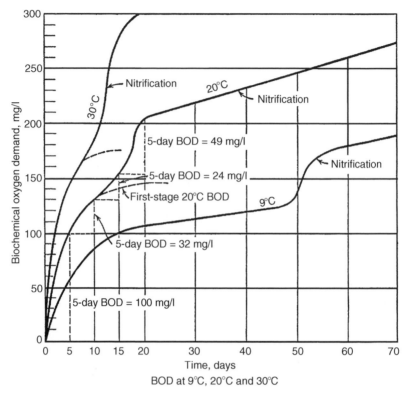

FIGURE 2.1 Variation of BOD with temperature.

OXYGEN UPTAKE IN A STREAM—THE OXYGEN SAG EQUATION

The discharge of an oxygen depleting substance into a stream is not always harmful or permanently polluting. As we saw above, the discharge of a biodegradable substance into a stream stresses the oxygen levels. However, the stream re-aerates itself somewhat in proportion to the oxygen deficit.

The re-aeration of a stream is in proportion to the decrease of the oxygen levels from equilibrium. The re-aeration is in proportion to oxygen transfer from the air, turbulence, and temperature. The basic rate of change in the oxygen deficit of a stream is given by the Streeter–Phelps equation. (U.S. Public Health Service Bulletin 146 (1925)). The deficit D is measured as follows:

The base formulation for the Oxygen Sag curve is as follows:

$$\mathrm{d}D/\mathrm{d}t = k_1 L - r_2 D$$

where D = Reference distance; D_a = the point of pollution or reference; k = BOD oxidation constant; r = rate of re-aeration; L_a = first stage BOD or BOD$_5$

$$D = \frac{kL_a(e^{-kt} - e^{-rt})}{(r - k)} + D_a e^{-rt}$$

When the equation is re-written, the following points of minimum oxygen concentration and inflection are found by the following equations:

$$f = r/k$$

$$D_c = (L_a e^{-kt})/f \quad \text{and} \quad t_c = [1/(k(f - 1))] \ln \{f[1 - (f - 1)(D_a/L_a)]\}$$

where D_c is the time to the critical oxygen level and t_c is the distance to the critical oxygen level.

The relationships are shown in Figure 2.2.

BIOLOGY OF POLLUTED WATER

The change in oxygen levels in the stream leads to changes in the aquatic environment. Many of these changes are reversible, but some are not. The changes occur not only to the chemistry but also to the biology and ecology of the stream.

As the oxygen levels start to drop, species begin to disappear. Certain species such as trout and game fish require a minimum dissolved oxygen content. If the oxygen content falls below the critical level, the fish cannot traverse the region, and migration for spawning is effectively eliminated.

As the oxygen content falls further and drops toward zero, the biota and the plants change. Benthic deposits develop in the stream and some metals begin to precipitate while others are reduced. The nitrates are reduced to ammonia (causing toxicity) and then further to nitrogen gas. When the nitrates are gone, the phosphates and the sulfates reduce next. By this time the oxidation reduction potential (ORP) is in the negative range, on the order of -100 mV or less. At this point, the river is a reducing environment, and the sulfates are reduced to H_2S, and dissolved metals are precipitated. Benthic deposits form on the river bottom, and the river water turns dark gray to black. The release of H_2S and some excess NH_3 is continuous, causing the "rotten egg" odor associated with anaerobic conditions.

The river is essentially "dead" until the carbon is consumed to the point where re-aeration from the surface can begin to supply oxygen to the river or until entering streams carrying dissolved oxygen have sufficient dilution to change the anaerobic conditions. At that point, the river can start to recover, but the ecology has changed. Eventually recovery may be complete; with the exception of the diversity of the species, there will be fewer species in the recovered downstream waters.

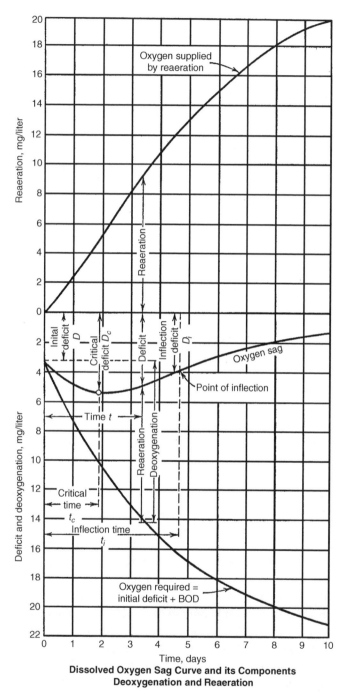

FIGURE 2.2   Variation in oxygen levels in a stream illustrating oxygen sag. *Source:* Fair and Geyer: Elements of Water Supply and Wastewater Disposal. John Wiley & Sons, 1963.

Biological Changes:

Loss of aquatic species

Development of anaerobic deposits further degrades water

Water turns black

Chemical reduction of metal oxides (iron and magnesium, and manganese)—changes in toxicity levels and solubilities of most metals due to the reducing environment

Development of benthic organisms and sludge worms

Total depletion of oxygen from nitrates and then from sulfates

High amine levels and ultimate release of N_2 as gas

Development of hydrogen sulfide and anaerobic conditions

Some metals solubilize (some re-precipitated)

Extremely slow stream recovery

Toxicity conditions that may persist long after recovery

Development of gradual recovery, but loss of habitat and some recovery may never occur until the pollution source is removed

A chart of the changes in the biota for polluted water is shown in Figure 2.3.

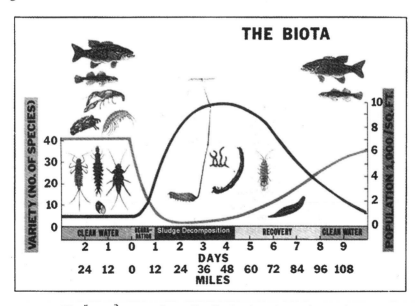

The [upper] curve shows the fluctuations in numbers of species: the [lower] the variations in numbers of each.

FIGURE 2.3 Variation in aquatic life when an oxygen sag occurs. *Source*: Public Works VSO (1959) pp. 104–110 by Bartsch and Ingram.

3

FLOW MEASUREMENT

Review of open channel hydraulics
Determining normal and gradually varied flows
Types of flowmeters
Weir plates

REVIEW OF OPEN CHANNEL HYDRAULICS

$$\text{Bernoulli equation:} \quad h = z + \alpha V^2/2g + p/\gamma$$

where h = hydraulic head or elevation of theoretical free surface; z = elevation; α = coefficient for velocity term; $V^2/2g$ = velocity squared divided by acceleration of gravity; p/γ = pressure/liquid weight (i.e., PSF/(lb/ft^3)).

Most open channel applications are *turbulent flow*. Although it is highly unlikely that laminar flow (Reynolds number under 2000) will be encountered, if the fluid is viscous, nonwater based, or contains extremely high levels of emulsified oils and other solids that may in fact be the case. If the channel or sewer has a distinct oil layer, that is a two phase flow regime, it is often appropriate to ignore the viscous layer unless there is clear evidence that the oil or viscous fluid has influenced the flow regimen. If so, then two phase analysis may be appropriate. Paper plant sewers may be particularly susceptible to this type of flow from tall oil sewers.

The Bernoulli equation and the Manning equation are used to define flow in open channels.

Practical Wastewater Treatment, by David L. Russell
Copyright © 2006 John Wiley & Sons, Inc.

The Manning equation has two forms: Metric and English. The difference is in the constant.

The Manning equation:
$$V = \frac{K\,R^{(2/3)}\,S^{(1/2)}}{n}$$

where V = velocity in an open channel due to gravity flow depending upon units; S = slope of the channel in ft/ft or m/m; R = hydraulic radius = wetted area/wetted perimeter = A/P; n = channel roughness coefficient, can be a variable, often selected from tabular values; K = constant for conversion = 1.486 for ft^3/s and English units and K = 1 for metric units.

This is the standard flow equation for all open channels and needs to be considered whenever gravity and friction forces predominate in the flow regime. Fortunately, this is true in about 95% of the open channel flow cases encountered.

Since many of our collectors are pipe systems, use the following graph to determine flows in sewers.

This is an aide to computing hydraulic radius. It will come in use later on. The following chart (Fig. 3.1) is most useful for circular sewers:

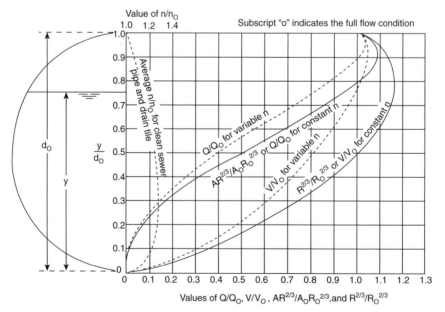

FIGURE 3.1   Flow characteristics of a circular sewer. *Source*: TR Camp, "Design of Sewers to Facilitate Flow", Sewage Works Journal, 1946, Vol. 18, No. 3.

Geometric Elements of Channel Sections

Section	Area A	Wetted perimeter P	Hydraulic radius R	Top width T	Hydraulic depth D
Rectangle	by	$b + 2y$	$\dfrac{by}{b+2y}$	b	y
Trapezoid	$(b+zy)y$	$b + 2y\sqrt{1+z^2}$	$\dfrac{(b+zy)y}{b+2y\sqrt{1+z^2}}$	$b + 2zy$	$\dfrac{(b+zy)y}{b+2zy}$
Triangle	zy^2	$2y\sqrt{1+z^2}$	$\dfrac{zy}{2\sqrt{1+z^2}}$	$2zy$	$\tfrac{1}{2}y$
Circle	$\tfrac{1}{8}(\theta - \sin\theta)d_0^2$	$\tfrac{1}{2}\theta d_0$	$\tfrac{1}{4}\left(1 - \dfrac{\sin\theta}{\theta}\right)d_0$	$\dfrac{(\sin\tfrac{1}{2}\theta)d_0}{2\sqrt{y(d_0 - y)}}$	$\tfrac{1}{8}\left(\dfrac{\theta - \sin\theta}{\sin\tfrac{1}{2}\theta}\right)d_0$
Parabola	$\tfrac{2}{3}Ty$	$T + \dfrac{8y^{2*}}{3\,T}$	$\dfrac{2T^2y^*}{3T^2 + 8y^2}$	$\dfrac{3A}{2y}$	$\tfrac{2}{3}y$
Round-cornered Rectangle $(y > r)$	$(\tfrac{\pi}{2} - 2)r^2 + (b+2r)y$	$(\pi - 2)r + b + 2y$	$\dfrac{(\pi/2 - 2)r^2 + (b+2r)y}{(\pi - 2)r + b + 2y}$	$b + 2r$	$\dfrac{(\pi/2 - 2)r^2}{b + 2r} + y$
Round-bottomed Triangle	$\dfrac{T^2r^2}{4z\,z}\left(1 - z\cot^{-1}z\right)$	$\dfrac{T}{z}\sqrt{1+z^2} - \dfrac{2r}{z}$ $\times\left(1 - z\cot^{-1}z\right)$	$\dfrac{A}{P}$	$2[z(y - r) + r\sqrt{1+z^2}]$	$\dfrac{A}{T}$

*Satisfactory approximation for the interval $0 < x \leq 1$, where $x = 4y/T$. When $x > 1$, use the exact expression $P = (T/2)[\sqrt{1 + x^2} + 1/x\,\ln(x + \sqrt{1 + x^2}\,)]$

49

Given the slope of the sewer and the approximate roughness (estimated from tables of materials of construction), one can approximate its discharge within 10% or so by using the formulas given above and the graphical approximation of the velocity and discharge percentage of the full pipe. The principal caution here is in the recognition of normal flow and in the ability to assign an appropriate value of Manning's n for the channel. By mathematical substitution it is possible to equate both the Colebrook formula coefficient and the f coefficients for pipe flow (from *Transactions ASME* Vol. 68, 1944, p. 627—commonly known as a Fanning Diagram) friction tables for confined flow to the Manning's n. Rapid or shooting flow and various flow regimens are discussed in the following sections.

For most flow computations one can largely ignore the Froude number in the computations but in flow measurement, including the installation of weirs and flumes, the Froude number for the approaching stream is very important. All open channel flowmeters operate on the principle of critical flow, where the Froude number is equal to 1, and there is a definite and reliable relationship between the flow through the flowmeter and the depth of flow in the flowmeter. If the approaching flow is less than $F = 1$, the measurements are reliable through the flowmeter, if however the value of F is greater than 1, the flow is unstable and the relationship between depth of flow and discharge is tenuous and the flowmeter will read low by significant amounts.

$F =$ Froude number – the point where gravity forces and friction forces are balanced

$$F = V/(\text{sqrt}(g \times L)) \text{ or}$$
$$F = V/(g \times D)^{1/2}$$

At flows having a Froude number greater than one, the flow is said to be supercritical or shooting flow. Any sudden disturbance that changes the flow depth (such as a weir plate) will cause the flow to jump to a "conjugate" depth or alternate depth based upon the Froude number. If the Froude number of a specific location (F_1 calculated at the upstream point) is equal to 1.7 or greater, then an undular hydraulic jump occurs. The undular jump is not steady in location or elevation and is often characterized by a variable downstream height (where F_2 is calculated), as a wavy surface profile with the jump face moving back and forth, often confused with a large ripple in a stream or natural channel. A strong jump (very steep vertical face) occurs when $F_1 > 9$; a steady jump occurs when $4.5 < F_1 < 9$;

an oscillating jump occurs when $2.5 < F_1 < 4.5$; a weak jump occurs when $1.7 < F_1 < 2.5$.

The conjugate depth (height of the free surface after the hydraulic jump) can be predicted by calculation. The conjugate depth of a hydraulic jump can be determined by solving the energy (Bernoulli) equation and plotting the depth of flow divided by the critical depth on a vertical axis, and the energy head divided by the critical depth on the horizontal axis. The resulting graph is a horizontal parabola with asymptotes at the horizontal axis and at a $45°$ angle from the origin, with the inflection point of the curve at 1.7. There are both simple and complex computer programs for computing these depths, or if it is important, the flow can be computed by any number of formulas given on the World Wide Web. For example, the Web sites http://onlinechannel.sdsu.edu/onlinechannel18.php, http://www.du.edu/~jcalvert/tech/fluids/opench.htm, and http://www.csus.edu/indiv/h/hollandm/ce135/HydrJump/HJVeiwGr.htm will give examples of calculation.

The important thing to recognize about the presence of a hydraulic jump is that the sudden change of depth can play havoc with measurements of flow and render the flowmeter highly inaccurate. An example of this was found at a large brewery in the southeast. The water from the above-ground clarifiers was allowed to drop about 10 m into the sanitary sewer. The water velocity was 4 m/s. From there it ran through one horizontal bend and then approximately 15 m into a broad crested weir flowmeter. The water was moving so fast that the flowmeter's depth–discharge relationship was unusable—for a very small change in measurable head there was a very large change in discharge. The plant engineering staff had other ways of calculating the plant water use, and finally abandoned the flowmeter, leaving it in place but no longer using it for effluent reporting.

A second example occurred when the effluent of a very large chemical complex was discharged into a 15% grade outfall into the Ohio River. The flow in the channel quickly became supercritical and had F values approaching 10. At this point, a small change in the depth of flow represents a very large change in the discharge in the effluent. Compounding this was the fact that the instrumentation engineer chose to select a flowmeter of low sensitivity, which was also subject to weather. The instrument, an electronic plumb bob on a string, raised and lowered by an automatic winch, had a sensitivity of about 0.1 cm (0.04 in.), which represented a change in flow of approximately 10%. The readings were extremely unreliable because the velocity of the water combined with wind on the cable prevented the accurate measurement of the water level surface.

DETERMINING NORMAL AND GRADUALLY VARIED FLOWS

In any channel of any shape there is a relationship between the channel slope and the normal flow. That is given by the Manning formula, shown in the first section. The normal depth (d_n) for the channel may be above or below the critical depth. The critical depth (d_c) is depth at which gravity and friction forces balance and $F = 1$.

Applying the Froude number when $F = 1$, yields the formula for critical depth d_c

$$d_c = V^2/g$$

Applying the Manning equation yields:

$$d_c = 1/n^2 \times 1/g \times R^{4/3} \times S,$$

and when there is a rectangular channel, one can make substitutions to express A/P as a function of d, and when one makes the substitution of $Q = V \times A,$

$$\text{the overall function is } d_c = \frac{Q^2 S}{g \times n^2 \times A^{2/3} \times P^{4/3}}$$

It is possible to combine the above equation much more as a function of the channel geometry and simplify the geometry so that d_c is a function of channel slope S and Q^2.

For any one value of Q there is one and only one value of d_c. This is why critical flow measurement devices work. This includes all types of flowmeters from Parshall flumes to Cutthroat flumes to Palmer-Bowlus flumes, and all types of weirs.

The flow in any open channel may be above or below the critical flow. The slope of the channel is classified as mild, critical, horizontal, steep, and adverse. As the liquid moves down the channel it will either increase or decrease in depth with distance. The behavior of the liquid in a channel is determined by the slope, channel roughness, and the flow. The depth of flow in the channel at any one point in time may be calculated and the flow depth at almost any other point predicted if one includes a consideration of the depth of the critical flow in the computations.

The flow varies in depth until it reaches certain asymptotes. Whenever the flow is below the critical depth, it will, when it encounters an obstacle, jump to a conjugate depth greater than the critical flow. This is shown in Figures 3.2 and 3.3.

When you have a surface where the flow is rapid or changing with depth, you cannot accurately measure it with an open channel device.

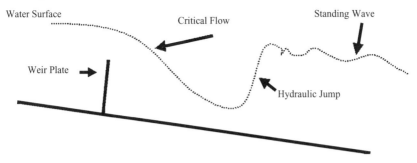

FIGURE 3.2 Elements of a hydraulic jump.

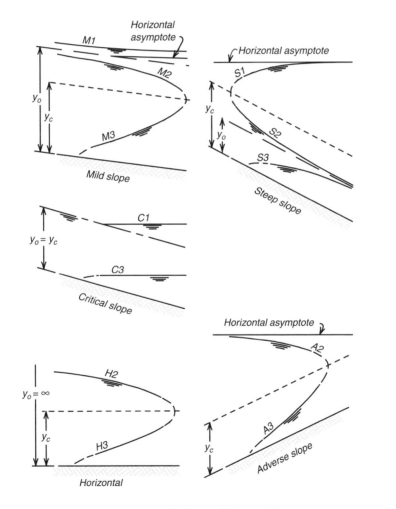

Surface profiles (backwater curves) for gradually varied flow
Flow is from left to right; vertical scales are greatly enlarged.

FIGURE 3.3 Surface profiles for different sloped channels curves and channel slopes are greatly enlarged.

TYPES OF FLOWMETERS

There are several types of flowmeters. The basic ones you will encounter are the following:

Parshall shapes and their variations
V-notch weirs
Broad crested weirs
Special shapes

The Parshall flume is shown in Table 3.1. It is a critical flow device. It must have free outflow downstream if it is to measure accurately. It must also have a reasonably quiescent upstream flow without sharp bends or corners for at least 6 weir lengths. It does handle solids well and is accurate over a reasonably broad range of flows. When installed properly and calibrated and maintained, it can be of primary standard quality. These types of flumes are suitable for flows containing solids and are self-cleaning. The Parshall flume is one of the most studied and the most widely used shapes for flow measurement. The geometry of the Parshall shape is critical and cannot be modified if the flume is to be used for accurate measurements. Accumulation of solids or films that might change the dimensions should also be avoided.

The Parshall flume and several other popular types, which are not subject to impairment by solids buildup, are shown in Table 3.1.

The Palmer-Bowlus and the Cutthroat flumes are also critical flow devices, varying only slightly in design and dimensions. The Palmer-Bowlus is often built as a slip in form for circular pipes, and it is generally accurate within 1–10% depending upon installation conditions, flow, surges, channel slope, and other factors. The Cutthroat flume is often used in rectangular sewers such as industrial plant drainage channels and is very widely used in agricultural applications. It has the advantage of having a flat bottom so there is no accumulation of solids.

For high accuracy, the flow should be measured by traversing the section using a portable flowmeter such as an ultrasonic flowmeter (typical unit made by Marsh-McBirney) or a Price current meter. Both can be used as "Primary Standard" devices in measurement of water flows. The Price current meter is the accepted standard device, and it has the ability to average flow over a few seconds or provide an "instantaneous" flow reading. The advertised accuracy of the Price meter is 2%. The principal obstacles to the Price current meter are the jeweled pivot, which the rotating cups rest upon, and the need for maintenance of these pins.

The Marsh-McBirney unit is also excellent and has an average time constant for ease of use, but it is dependent upon and can be affected by conductivity. In a chemical plant, such as a paper plant, the Marsh-McBirney

TABLE 3.1

a. Plan and sectional elevation of concrete Parshall measuring flume

b. Palmer Bowlus flume

c. Cutthroat flume

$B = W + 2L_1/3 = W + L_2/3$
$L_a = 2L/9$
$L_b = 5L/9$

W	A	Z	B	C	D	E	F	G	K	N	R	M	P	X	Y	Free-Flow Capacity Minimum	Free-Flow Capacity Maximum
Ft. In.	Ft. In.	Ft. In.	Ft. In.	Ft. In.	Ft. In.	Ft.	Ft.	Ft.	In.	In.	Ft. In.	Ft. In.	Ft. In.	In.	In.	cfs	cfs
0-3	1-6.3/8	1-1/4	1-6	0-7	0-10.3/16	2	1/2	1	1	2.1/4	1-4	1-0	2-6.1/4	1	1.1/2	0.03	1.9
0-6	2-7/16	1-4.5/16	2-0	1-3.1/2	1-3.5/8	2	1	2	3	4.1/2	1-4	1-0	2-11.1/2	2	3	0.05	3.9
0-9	2-10-5/8	1-11.1/8	2-10	1-3	1-10.5/8	2.1/2	1	1.1/2	3	4.1/2	1-4	1-0	3-6.1/2	2	3	0.09	8.9
1-0	4-6	3-0	4-4.7/8	2-0	2-9.1/4	3	2	3	3	9	1-8	1-3	4-10.3/4	2	3	0.11	16.1
1-6	4-9	3-2	4-7.7/8	2-6	3-4.3/8	3	2	3	3	9	1-8	1-3	5-6	2	3	0.15	24.6
2-0	5-0	3-4	4-10.7/8	3-0	3-11.1/2	3	2	3	3	9	1-8	1-3	6-1	2	3	0.42	33.1
3-0	5-6	3-8	5-4.3/4	4-0	5-1.7/8	3	2	3	3	9	1-8	1-3	7-3.1/2	2	3	0.61	50.4
4-0	6-0	4-0	5-10.5/8	5-0	6-4.1/4	3	2	3	3	9	1-8	1-6	8-10.3/4	2	3	1.3	67.9
5-0	6-6	4-4	6-4.1/2	6-0	7-6.5/8	3	2	3	3	9	2-0	1-6	10-1.1/4	2	3	1.6	85.6
6-0	7-0	4-8	6-10.3/8	7-0	8-9	3	2	3	3	9	2-0	1-6	11-3.1/2	2	3	2.6	103.5
7-0	7-6	5-0	7-4.1/4	8-0	9-11.3/8	3	2	3	3	9	2-0	1-6	12-6	2	3	3.0	121.4
8-0	8-0	5-4	7-10.1/8	9-0	11-1.3/4	3	2	3	3	9	2-0	1-6	13-8.1/4	2	3	3.5	139.5
10	14-3.1/4	6-0	14	12-0	15-7.1/4	4	3	6	6	13.1/2	8-0	6-0	32-0	12	9	6	200
12	16-3.3/4	6-8	16	14-8	18-4.3/4	5	3	8	6	13.1/2	9-0	8-0	35-0	12	9	8	350
15	25-6	7-8	25	18-4	25-0	6	4	10	9	18	11-0	9-0	40-0	12	9	8	600
20	25-6	9-4	25	24-0	30-0	7	6	12	9	27	12-0	10-0	48-0	12	9	10	1,000
25	25-6	11-0	25	29-4	35-0	7	6	13	12	27	12-0	10-0	55-0	12	9	15	1,200
30	26-6.3/4	12-8	26	34-8	40-4.3/4	7	6	14	12	27	13-0	10-0	64-0	12	9	15	1,500
40	27-7.1/2	16-0	27	45-4	50-9.1/2	7	6	16	12	27	13-0	11-0	80-0	12	9	20	2,000
50	27-7.1/2	19-4	27	56-8	60-9.1/2	7	6	20	12	27	13-0	11-0	95-0	12	9	25	3,000

Parshall flume, with dimensions for various throat widths, and other common flume types

55

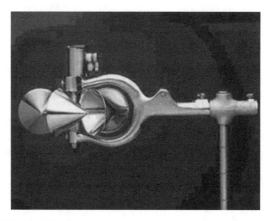

FIGURE 3.4   Pygmy current meter by Gurley instruments.

meter is unreliable when measuring flows from "green liquor" or other areas where the liquid has little or no conductivity. The Marsh-McBirney unit also has a higher demand on the batteries. The Price and units is shown in Figure 3.4.

WEIR PLATES

For routine measurement applications where there are little or no solids in the flows, and where the flow does not vary widely, it is difficult to beat a weir plate measurement device for both installation cost and accuracy. Several of the weir plates and their equations are illustrated below. The most popular of these are the rectangular weirs and the v-notch weirs. In all cases except the proportional or Sutro weir shown below, the water level is measured a bit upstream from the weir plate with the level of the lowest point on the plate as a reference.

Installation conditions, such as setting the weir plates at an angle to the flow, or failure to account for the approach velocity can cause both erratic and unreliable measurements. The failure to account for the velocity of the flow in the approach channel will cause the weir to measure between 10% and 15% low.

More detailed information on flow measurement can be obtained from the USGS Water Measurement Manual, by conducting a World Wide Web Search or by logging on to http://www.usgs.gov. The book is a classic and covers a wide variety of flow measurement discussions. The book is formatted in Adobe Acrobat®.

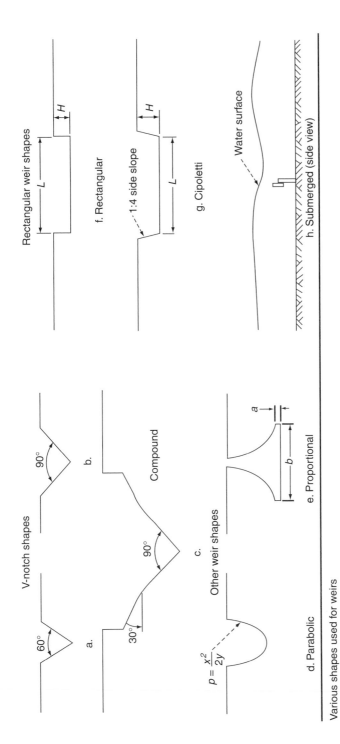

FIGURE 3.5 Various standard weir shapes.

TABLE 3.2   Discharge Equations for Common Weir Shapes

Name	Discharge Equation	Comments
60° V-notch	$Q = 1.43\ H^{2.5}$	Approximate formula
90° V-notch	$Q = 2.49\ H^{2.48}$	
Combination	$Q = 3.9\ H^{1.72} - 1.5 + 3.3\ Lh^{1.5}$	
Parabolic	$Q = 1.512 p^{0.478} h^2$	$Q \propto h^2$
Proportional	$Q = Ca^{1/2}\ b\sqrt{2g}(h - \frac{2}{3})$	$0.625 < C < 0.600\ Q \propto t$
Rectangular	$Q = 3.33\ H^{3/2}\ (L - 0.2H)$	Fully contracted ends
Cipolletti	$Q = 3.367\ LH^{1.5}$	
Submerged	$Q = 3.33\ L\ (n\ H)^{3/2}$	$n =$ tabular correction for submergence ratio.

The Sutro or proportional weir is a specialty shape. It is designed to have the flow directly proportional to the depth. This type of weir is subject to the same limitations as other weirs, but it is often found in grit removal chambers of sewage works. The basis for their application is that they will provide a constant horizontal velocity through the approach channel, allowing removal of various types of grit and sand particles from the liquid.

The formula for Sutro weirs is as follows:

$$x/b = 1 - (2/\pi)\tan^{-1}\sqrt{y/a}$$

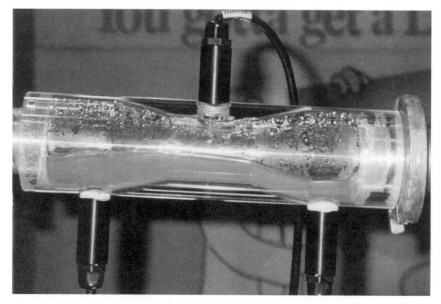

FIGURE 3.6 A modified Venturi/ Parshall shape acting as an Open Channel Flowmeter Sutro weirs with ultrasonic depth monitor.

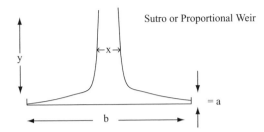

FIGURE 3.7   Proportional or Sutro weir shape and dimensions.

The discharge equation is $Q = \{Ca^{1/2}b\sqrt{2g}(h - a/3)\}$

The drawings for Table 3.1, and Figures 3.5 & 3.7 were reproduced from an article by the author, with permission from *Chemical Engineering* Magazine, October 20, 1980, pp. 109–121.

4

SAMPLING AND STATISTICAL CONSIDERATIONS

Errors in process measurements
Statistical distributions
Lognormal distributions
Weibull distributions
Probable error
Repeat measurements
Sampling

ERRORS IN PROCESS MEASUREMENTS

In any set of environmental measurements, the subjects of accuracy and precision of the measurements are always beneath the surface. "How good are the measurements?" The difference between accuracy and precision is shown in Figure 4.1.

There is a regulatory problem in the United States. Most environmental discharge permits embody normally distributed statistics for environmental events. This is incorrect. Most environmental data are distributed either lognormally or in accordance with a Weibull type III distribution. The question of why we should not use a standard distribution is simply explained by the examination of the distribution curves (See Fig. 4.2.)

By looking at the above distributions, one can see that the mean and the average are not the same value. This is generally true in almost all

Practical Wastewater Treatment, by David L. Russell
Copyright © 2006 John Wiley & Sons, Inc.

Precision versus Accuracy

High Precision, Low Accuracy   High Accuracy, Low Precision

The objective was to score a bullseye with the target. The left target shots are highly precise but inaccurate. The right target shots are imprecise but their average is within the bullseye

FIGURE 4.1   The difference between Precision and Accuracy.

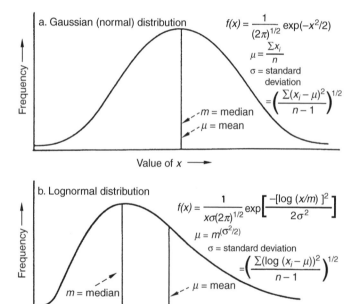

a. Gaussian (normal) distribution

$$f(x) = \frac{1}{(2\pi)^{1/2}} \exp(-x^2/2)$$

$$\mu = \frac{\Sigma x_i}{n}$$

σ = standard deviation

$$= \left(\frac{\Sigma(x_i - \mu)^2}{n-1}\right)^{1/2}$$

m = median
μ = mean

Value of x

b. Lognormal distribution

$$f(x) = \frac{1}{x\sigma(2\pi)^{1/2}} \exp\left[\frac{-[\log (x/m)]^2}{2\sigma^2}\right]$$

$$\mu = m^{(\sigma^2/2)}$$

σ = standard deviation

$$= \left(\frac{\Sigma(\log (x_i - \mu))^2}{n-1}\right)^{1/2}$$

m = median μ = mean

Value of x

Normal distributions

FIGURE 4.2   Normal and lognormal distributions. *Source:* "Monitoring and Sampling Liquid Effluents," by D. L. Russell, PE, Chemical Engineering, October 20, 1980[1]

[1]Russell DL, PE. Monitoring and Sampling Liquid Effluents. Chemical Engineering, October 20, 1980.

environmental data. If we want to apply a safety factor around the mean value to incorporate the highest probable value, we use multiple values of the standard deviation. In a lognormal distribution it is impossible to have a negative value, just as it is in the environment.

Example: Let us assume that we are applying for a water discharge permit. The discharge permit is written in terms of mass limits, expressed both in pounds and kilograms per day. Generally, most permit writers use a normal or Student's distribution despite the fact that the data are not normally distributed. This can cause significant problems for most facilities. With reference to the data below, let us assume that the mean in the data is 100 kg/day, and the standard deviation is 60 kg/day. Given a specific distribution, most EPA permit writers would express the permit conditions in terms of both a 30-day average and a daily maximum. Most permit writers will express the upper limit in the permit in a manner to include at least 95% of the possible values and express the upper limit or daily maximum value as the daily average plus two times the standard deviation. In our case that would give us a daily average of 100 kg/day and a daily maximum of $100 + 60 \times 2 = 220$ kg/day. According to standard statistical theory, the area under the normal curve will contain 96% of the values when a two standard deviation allowance is used.

Look at the curve carefully. When you apply a two standard deviation on each side of the mean, it indicates that the values expected can be anywhere from -20 kg/day to 220 kg/day, and the implicit assumption is that when you have a really good day you might reach a value of zero or less. That is patently impossible. Industries and facilities do not work that way.

Now examine the lognormal distribution. Because it is skewed to the right and has a heavy tail, in order to have a 95% + chance of having your permit encompass all the possible effluent values, two things are required: Your lower permit limit would be greater than zero, and your upper permit limit would be significantly greater than the 220 kg/day figure.

It has been only recently that the permit writing branch of the USEPA has acknowledged that the permit writer may not necessarily have to use the standard deviation in permit preparation, and special allowance have now found their way into the permit writer's guide, but it is still too easy for the unchallenged or uninitiated permit writer to develop a permit based on standard statistics, sometimes with disastrous consequences.[2]

---

[2]One facility had such a problem, and it is described, in general terms, in an article in Chemical Engineering, October 9, 1978, "Measurement Uncertainties in the NPDES Permit System," by D.L. Russell and J.J. Tiede, p. 115.

STATISTICAL DISTRIBUTIONS

The most common distribution is the normal distribution. Unfortunately, it is about the only distribution understood by the regulatory community.

Most environmental data have either a lognormal or a Weibull type III distribution.

The basic problem is the high values. In environmental systems these high values are real. The lognormal and Weibull distributions include them and can give a reasonably good fit for data. In normal statistical distributions these values are often treated as outliers and can lead to permit violations and fines.

LOGNORMAL DISTRIBUTIONS

A lognormal distribution is one where the following occurs:

$$\bar{c}f(x) = \frac{1}{x\sigma_y\sqrt{2\pi}} \exp\left[-\frac{1}{2\sigma_y^2}(\ln x - \mu_y)^2\right] \quad x > 0, \quad -\infty < \mu_y < \infty, \quad \sigma_y > 0$$

where μ_y and σ_y are the true mean and variance of the transformed variable $Y = \ln X$

The thing that characterizes lognormal distributions is that there are *no negative numbers in the distribution. This is also true of environmental data.*

Richard O. Gilbert suggests using a three parameter lognormal density function with the value τ because it allows us to shift the axis without changing the shape of the distribution.[3] The revised equation is as follows:

$$f(x) = \frac{1}{(x-\tau)\sigma_y\sqrt{2\pi}} \exp\left\{-\frac{1}{2\sigma_y^2}\left[\ln(x-\tau) - \mu_y\right]^2\right\}$$

$$x > \tau, -\infty < \mu_y < \infty, \sigma_y > 0, -\infty < \tau < \infty$$

and the distributions are shown in Figure 4.3.

---

[3]Gilbert, Richard O. Statistical Methods in Environmental Monitoring, Van Nostrand Rinhold, 1987. This is the statistical book that I would consider purchasing if I did not already own it.

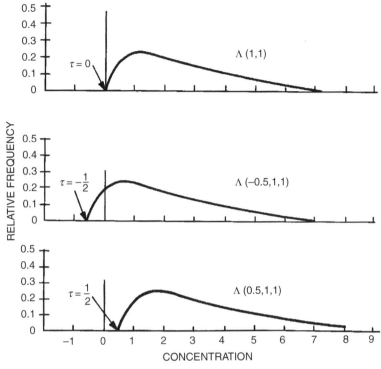

FIGURE 4.3   Three parameter lognormal density distribution.

The transformation can help us account for average values and for the extremes we so often encounter.

The computation of various parameters used in the estimation of goodness of fit and other measures is beyond the scope of this text and the reader is referred to Goodman for that work. (See footnote 3)

WEIBULL DISTRIBUTIONS

This is one of the other common distributions. Meteorological data are estimated using a Weibull type III distribution. That enables one to estimate the size of a given precipitation event and the probability that it will occur with a given frequency.[4] The Weibull distribution is shown in Figure 4.4.

---

[4]Technical Publication No. 40—The Rainfall Frequency Atlas of the United States is a reference work on rainfall probability and return rate for the eastern US. The Rainfall Frequency Atlas (TP-40) is available at http://srh.noaa.gov/lub/wx/precip_freq/precip_ index.html. The Atlas for the western US can be found at: http://www.wrcc.dri.edu/ CLIMATEDATA.html

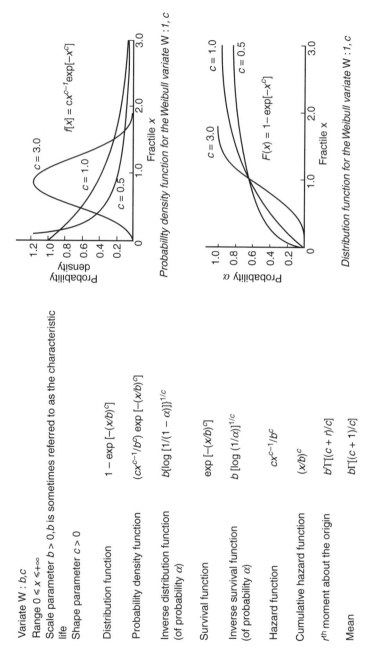

Variate W : b,c

Range 0 ⩽ x ⩽ +∞

Scale parameter $b > 0$, b is sometimes referred to as the characteristic life

Shape parameter $c > 0$

Distribution function	$1 - \exp[-(x/b)^c]$
Probability density function	$(cx^{c-1}/b^c) \exp[-(x/b)^c]$
Inverse distribution function (of probability α)	$b[\log[1/(1 - \alpha)]]^{1/c}$
Survival function	$\exp[-(x/b)^c]$
Inverse survival function (of probability α)	$b[\log(1/\alpha)]^{1/c}$
Hazard function	cx^{c-1}/b^c
Cumulative hazard function	$(x/b)^c$
rth moment about the origin	$b^r\Gamma[(c + r)/c]$
Mean	$b\Gamma[(c + 1)/c]$

FIGURE 4.4 The Weibull distribution.

The distribution further resolves some of the extreme values; which are not addressed by the normal distribution.

PROBABLE ERROR

In order to be able to measure the uncertainty in our measurement, we must understand the concept of probable error.

The probable error in a related measurement (M) is given by the following. If a measurement is composed of individual and independent functions, $f_1, \ldots, f_n$

$$M = f_1(x) + f_2(x) + f_3(x) + \cdots$$

The error in M is given by the following

$$e^2 = (e_1 dx/df_1)^2 + (e_2 dx/df_2)^2 + \cdots$$

where e is the error of measurement in the individual parameter.

This type of error estimation can be used in all measurements and equations, even air pollution stack tests, and complex discharge measurements as well as measurement errors for property surveys.

Consider the following example as a problem set:

You have permit that is issued in mass units – that is, kg/day. The flowmeter is a $0.3937M$ (1 ft) Parshall flume with a discharge equation of Q (CuM/D) $= 59688.0\,H^{1.522}$ where H is in meters.

The average discharge of suspended solids is 50 mg/l and the published figure for the accuracy of the test is 15%. The uncertainty in the measurement of depth (due to surface waves) is 0.01 M (0.3937 In.)

The average suspended solids discharge is:

$$0.05 \text{ g/M}^3 \times 15939.417 \text{ M}^3/\text{day} = 796.97085 \text{ g/day} = 0.797 \text{ kg/day}$$

How accurate is your measurement?

The permitted suspended solids load is 2.00 kg/day. How much over the permit limit are you? What is the uncertainty and how can you increase your accuracy?

In this case, the discharge is M (mass) $= C \times Q$

The error is given by:

Error in $H = 0.01$, Error in $C = 0.15 \times 50 = 7.5 \text{ mg/l} = 0.075 \text{ g/M}^3$

$$dM/dC = 59688 \times H^{1.522} \text{ and } dM/dH = 1.522 \times 59688 H^{0.522}$$

$$e^2 = (e_C \times dM/dC)^2 + (e_H \times dM/dH)^2$$
$$= [0.01 \times 1.522 \times 59688(0.42)^{0.522}] + [0.075 \times 59688(0.42)^{1.522}]^2$$
$$= 59688^2([0.01 \times 1.522 \times 0.42^{0.522}]^2 + [0.075(0.42)^{1.522}]^2)$$
$$= 59688^2 \times [0.00967^2 + 0.020028^2] = 59688^2 \times 0.01017$$

$e = 6019.87$ g/day $= 6.02$ kg/day, which is the probable error in the measurement of discharge. The problem is that the probable error is larger than the measured value by a factor of about 4. Therefore, you must increase the number of measurements in order to determine whether you are in compliance or noncompliance with your permit. If your mass discharge measurement was 10 kg or even 2.01 kg you would be recording a permit violation. But right now you do not know whether you are over or under your permit value.

REPEAT MEASUREMENTS

If you are going to take multiple measurements on the same source, what is the consequence?

If $C = A - B$

then the error in C is var (A) + var (B) − 2 covar (A,B)

and if A and B are independent then covar $(A,B) = 0$

Now for measurements where $C = A \times B$

var $(C) =$ var $(A \times B)$

which is approximated by the Taylor expansion as:

$$\text{var}(C) = (\delta C/\delta A)^2 B_\mu + (\delta C/\delta B)^2 C_\mu + \cdots,$$

which is the formula given above.

Now for net calculations, of the form $D = A_E B_E - A_I B_I$ where E is the effluent and I is the influent. Now if for the purposes of symbolism we allow X to represent the mean and S the variance at the mean, the following formula represents the net of the variance or error in the measurement of D:

Then $e^2{}_D = (X^2{}_{BE} \times S^2{}_{AE} + S^2{}_{AE} \times X^2{}_{BE}) + (X^2{}_{BI} \times S^2{}_{AI} + X^2{}_{AI} \times S^2{}_{BI})$

and for multiple analyses, the average error $= \sqrt{e/n}$, where n is the repeat number of analyses. All this assumes that we have a relatively constant process and it is not subject to variance within the plant from processes starting and stopping and variations in the influent levels of parameters. If any of those conditions occur, the calculations get much more messy.

Using our example above and assuming (with proper caution) that we have a normal statistical distribution, we can decrease the probable error by increasing the number of duplicate samples by $e_M = e_{(ave)}/(\text{No. of repeat measurements})^{0.5}$.

SAMPLING

A discussion on sampling is contained in the following articles, which appeared in *Chemical Engineering:*

"Monitoring and Sampling Liquid Effluents", October, 1980.

"Measurement Uncertainties in the NPDES Permit System", with J. Tiede, October, 1978.

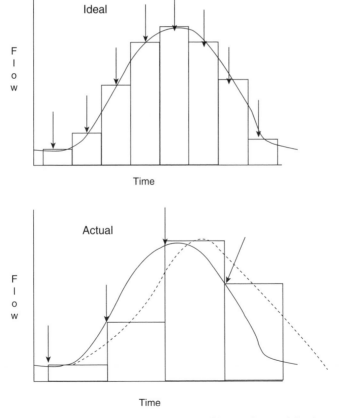

FIGURE 4.5   Ideal versus Actual sampling profiles on time weighted averages.

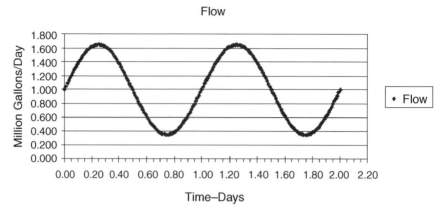

FIGURE 4.6   Typical diurnal variation in wastewater flows.

Consider for a minute the following flow curves (Fig. 4.5):

In the top figure the sampling is conducted at an "ideal" interval where we know something about the flow patterns in the next time interval. In real life we don't know what will happen next and the flow can just as easily spike as decline. The point at the lower figure is that the sampling program distorts the apparent data at least by one half of the sampling interval; and smooths out peaks & valleys and concentration fluctuations.

A typical municipal flow pattern is shown in Figure 4.6 expressed as $Q = Q_{AVG}(1 + 0.65 \sin (2\pi \cdot t))$ where t is in days. Sampling at every 2 hrs (0.0833d) will give one result. Sampling at 3 or 4 hours will give a different result.

5

IMPORTANT CONCEPTS FROM AQUATIC CHEMISTRY

Common ion species

Most important chemicals in the water environment

Carbonate chemistry

Chemical water softening

Excess lime process

Metals removal by precipitation

Heavy metals

Chromium reduction and metals precipitation

Silicates in treatment systems

Nitrogen

Sulfur

Phosphorous

In this chapter we will discuss each topic as it relates to the environment and waste treatment.

COMMON ION SPECIES[1]

The ion species discussed here are some of the most important inorganic compounds found in the aquatic environment. They are found abundantly in the Earth's crust. The relative importance of the ion species and the reaction

[1]Mathematical Modeling for Water Pollution Control Processes, Keniath & Wanielista, Ann Arbor Science, 1975.

Practical Wastewater Treatment, by David L. Russell
Copyright © 2006 John Wiley & Sons, Inc.

TABLE 5.1

Reaction	Equilibrium Constant	Log K
$H_2O = H^+ + OH^-$	K_w	-14.0
$H_2CO_3 = H^+ + HCO_3^{2-}$	K_1	-6.2
$HCO_3^- = H^+ + CO_3^{2-}$	K_2	-10.2
$H_3PO_4 = H^+ + H_2PO_4^-$	K_1	-2.2
$H_2PO_4^- = H^+ + HPO_4^{2-}$	K_2	-7.2
$HPO_4^{2-} = H^+ + PO_4^{3-}$	K_3	-12.2
$NH_4OH^- + H^+ = NH_4^+ + H_2O$	K	9.2
$Al^{3+} + OH^- = Al(OH)^{2+}$	$K_{(AlOH^{2+})}$	9.0
$Al^{3+} + 4(OH^-) = Al(OH)_4^-$	$K_{(Al(OH)_4^-)}$	32.5
$Ca^{2+} + CO_3^{2-} = CaCO_3$	$K_{(CaCO_3)}$	3.2
$Ca^{2+} + HPO_4^{2-} = CaHPO_4$	$K_{(CaHPO_4)}$	2.7
$Ca^{2+} + PO_4^{3-} = CaPO_4^-$	$K_{(CaPO_4^-)}$	6.5
$Ca_5(PO_4)_3OH = 5Ca^{2+} + 3PO_4^{3-} + OH^-$		-49
$CaCO_3 = Ca^{2+} + CO_3^{2-}$		-8
$Al(OH)_3 = Al^{3+} + 3OH^-$		-30.4
$Al_{1.4}PO_4(OH)_{1.2} = 1.4Al^{3+} + PO_4^{3-} + 1.2\,OH^-$		-32.2

constant or stability constant is important because of the large number of reactions into which these chemicals participate. Carbonate chemistry and the reactions with calcium enter into most aquatic reactions, including saltwater reactions.

The reaction constants are also referred to as disassociation constants. They have the form of

$[A^+] + [B^-] = C$, where the constant is expressed as $Kc = [A^+][B^-]/[C]$ and the concentrations of the compounds are expressed in moles.

MOST IMPORTANT CHEMICALS IN THE WATER ENVIRONMENT

Most of the important aquatic inorganic chemistry involves a relatively limited set of cations—aluminum, calcium, iron, magnesium, manganese, phosphorous, and sodium, and a similar set of anions—chlorine, nitrogen, oxygen, and sulfur for the simple reason that these are the abundant elements. This does not preclude any of the many other reactions listed in the following sections, nor the importance of being able to control and manipulate the ions and ion species in the aquatic environment. Carbon chemistry was not included in this list because it is the foundation of organic chemistry, and there are libraries filled with organic chemistry reactions and interactions; this book could not begin to discuss adequately (Table 5.1).

CARBONATE CHEMISTRY

The chemistry of the carbonates and bicarbonates is important in the water treatment, both because it plays a role in the biological treatment and because a knowledge of it is also useful in dealing with various types of precipitation for industrial wastes and scale prevention for boiler waters. Some of the following concepts pre-date molar chemical theory, and to the modern student of chemistry, the calculation of equivalents of calcium carbonate and balancing equations using carbonate equivalents is archaic in view of the modern molar theories of chemical interactions. However, the system is still widely used and shows no signs of going away. It is used actively in the water treatment and water-softening industry.

The most common measure of the carbonate chemistry is the acidity and alkalinity tests.

Acidity is measured by the titration of water to the phenolphthalein endpoint. It is complete at about pH 8.5. All waters with a pH of less than 8.5 are assumed to have some acidity. Acidity due to CO_2 takes place between pH 4.5 and pH 8.5, and the phenolphthalein endpoint is between pH 8.2 and pH 8.4. Below pH 4.5 the acidity is considered as mineral acidity. The titration is performed with 0.02 N NaOH (N/50 NaOH).

The sources of acidity include the dissolution of carbon dioxide to create carbonic acid and the presence of other minerals. For carbonate acidity:

$$[H^+][HCO_3^-]/[H_2CO_3] = K_1 \approx 4.45 \times 10^{-7}$$

Mineral acidity in water is generally associated with any water that has a pH of less than 4.5. Titration of mineral acidity is assumed complete by the time the pH of 4.5 is reached.

Sources of mineral acidity include the oxidation of sulfur pyrites to sulfurous and sulfuric acids, and other mineral compounds to their equivalent acid forms. These acids are often found in drainage associated with coal mines and in the anthracite coal producing areas of Pennsylvania and West Virginia; it is not uncommon, even today, to find streams with a pH of less than 4.5 due to the presence of acid mine drainage caused by the oxidation of sulfur in coal formations.

$$2S + 31/2\,O_2 + 2H_2O \rightarrow 2\,H_2SO_4$$

Alkalinity is due to the presence of salts, principally in the form of bicarbonates and salts of weak acids. It is measured by titration with 0.02N H_2SO_4 (N/50 H_2SO_4). If the initial pH is above 8.3 the titration is done in two steps using phenolphthalein endpoint as a first indicator. Methyl orange is used as the second endpoint. The titration is performed with 0.02N H_2SO_4.

Water can have both acidity and alkalinity at the same time.

Graphical Representation of Alkalinity Titration

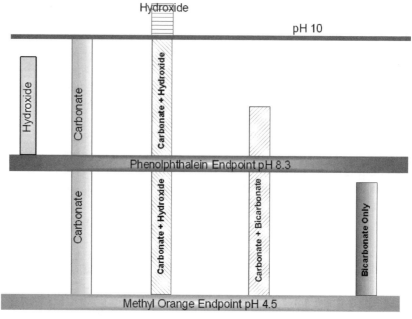

FIGURE 5.1   Graphical representation of alkalinity determination by titration.

Figure 5.1 shows a graphical representation of the way in which alkalinity is calculated.[2]

Hydroxide alkalinity gets consumed by the time the pH is 8.3, but carbonate alkalinity only gets half neutralized by that point.

The following characteristics are generally observed:

Samples having bicarbonate alkalinity have a pH between 8.3 and 11.

Samples having high hydroxide alkalinity have a pH usually above 10.

In a mixed sample having hydroxide and carbonate alkalinity, the correction for hydroxide alkalinity is: total alkalinity − carbonate alkalinity or

$$Alk_{OH} = Alk_{TOTAL} - 2\,Alk_{CARB}\,[\text{from pH 8.3 to pH 4.5}]$$

The relationship between the various species is both shown and described by the following chemical relationships:[3]

---

[2]Sawyer, Clair N. Chemistry for Sanitary Engineers. New York: McGraw Hill, 1960.
[3]From Sawyer, Op Cit.

When Temperature T is given in Kelvin ($^\circ C + 273.18$)

$$[H^+][OH^-] = K_w = 10^{-14}$$

$$[H^+][HCO_3^-]/[H_2CO_3] = K_1 = 10^{(14.8435-3404.71/T-0.032786T)}$$

$$[H^+][CO_3^{2-}]/[HCO_3^-] \quad K_2 = 10^{(6.498-2909.39/T-0.02379T)}$$

and $\qquad [Ca^{2+}][CO_3^{2-}] = K_s \quad 4.82 \times 10^{-9}$ approximately, the following relationships can be developed :

Looking at the overall equilibria, we can find the following relationships:
Total carbonate species, $Ct = [H_2CO_3] + [HCO_3] + [CO_3^{2-}]$

$$a_0 = [H_2CO_3]/Ct$$

$$a_1 = [HCO_3^-]/Ct$$

$$a_2 = [CO_3^{2-}]/Ct$$

and by a series of algebraic manipulations, the following relationships are developed:

$$a_0 = \frac{1}{1 + K_1/[H^+] + K_1K_2/[H^+]^2}$$

$$a_1 = \frac{1}{[H^+]/K_1 + 1 + K_2/[H^+]}$$

and $\qquad a_2 = \dfrac{1}{[H^+]^2/(K_1K_2) + [H^+]/K_2 + 1}$

and for a CO_2 saturated system $a_0 + a_1 + a_2 = 1$

However, as in most situations, where the water is at less than saturation point but is in equilibrium with the atmosphere, the following general relationships hold:

Electroneutrality must be satisfied, so that

$$[Cations] + [H^+] = [HCO_3^-] + 2[CO_3^{2-}] + [OH^-] + [Anions].$$

Alkalinity of the system $= [Z] = [C] - [A]$ and after some more appropriate manipulations, $[Z] = a1[Ct] + 2a_2[Ct] + K_w/[H^+] - [H^+]$

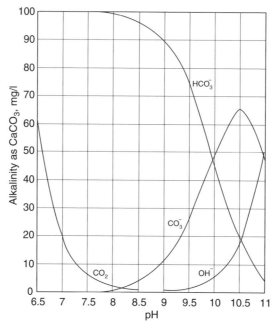

Relationship between carbon dioxide and the three forms of alkalinity at various pH levels. (Values calculated for a water with a total alkalinity of 100 mg/l at 25°C.)

FIGURE 5.2   Relationship between carbon dioxide and alkalinity.

and for limestone formations that contain groundwater, generally saturated with $CaCO_3$ but not with CO_2, $2[Ca^{2+}] + [Z'] + [H^+] = [HCO_3^-] + 2[CO_3^{2-}] + [OH^-]$

where $[Z']$ is alkalinity minus the calcium ion concentration.[4]

Using the relationships described above, one can determine the total carbonate species of water from a limestone formation by use of the pH alone. The quadratic formula can be used in the formula:

$$\underset{a}{\underbrace{\frac{(a_1 + 2a_2)}{a}}}(Ct)^2 - \underset{b}{\underbrace{\frac{([Z'] + [H^+] - K_w/[H^+])}{b}}}(Ct) - \underset{c}{\underbrace{\frac{2K_s/a_2}{c}}} = 0, \quad \text{which is the form}$$

$$Ct = \frac{-b \pm \sqrt{(b^2 - 4ac)}}{2a}$$

The use of the quadratic equation then permits calculation of the carbonate species (see Fig. 5.2).

[4]Material developed from Rich, Environmental Systems Engineering, McGraw Hill, 1960, and Water Quality and Treatment, 5th Edition, AWWA.

Alkalinity is necessary for denitrification in aerobic processes. Alkalinity is always measured in terms of $CaCO_3$. Biological nitrification and denitrification will be discussed later.

Hardness is often responsible for scale forming in water in cooling towers and in pipes. It is almost always caused by divalent metallic ions present in the water. To obtain a complete profile on the hardness, it is necessary to run a cation balance on the waters in question.

Hardness is classified in two ways, carbonate and noncarbonate hardness, and also classified with respect to the ions, calcium and magnesium. The hardness in water not chemically related to bicarbonates is noncarbonate hardness.

$$\text{Total hardness} = \text{Calcium hardness} + \text{Magnesium hardness} + \text{Noncarbonate hardness}$$

and

$$\text{Alkalinity (mg/l)} = \text{Carbonate hardness (mg/l)}$$

For each ion fraction

$$\text{Hardness (mg/l) as } CaCO_3 = M^{2+}(\text{mg/l}) \times 50/(\text{eq. wt of } M^{2+})$$

So that for iron the equivalent weight would be 55.845/2 (MW/ valence) or 27.9225 but we can approximate – and the multiplier would be 50/55.9 or 0.895.

Sometimes there is more alkalinity in the water than is necessary to satisfy the divalent cations. This is particularly true in alkaline waters. This is known as negative noncarbonate hardness and is associated with the presence of K^+ and Na^+ ions in the water. It is necessary to know the hardness of the water in domestic water treatment and in chemical precipitation as the latter often represents a specific chemical demand both in ion exchange and in chemical precipitation. It is also useful to know because in chemical precipitation, it is often easier to use chemical "water softening" techniques to assist in the removal of specific ions and other materials.

$$\text{Noncarbonate hardness (NCH)} = \text{Total hardness} - \text{Alkalinity}$$

The relationships between the forms are shown in Table 5.2.

The overall approach for analyzing a water includes first to diagram the ion balance in the water and then to make decisions about the way in which to treat the water to remove the excess ions.

TABLE 5.2   Hardness Relationship in Water

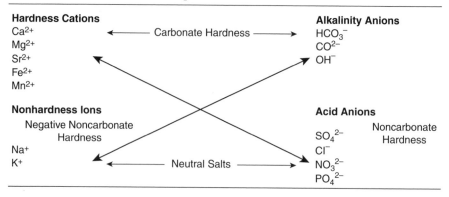

Example: Given the following analyses (see Table 5.3):

First, all alkalinity is bicarbonate because of pH (see Table 5.3).
   Next, calculate the $CaCO_3$ equivalents as shown below and construct a bar chart.

$$Ca = 2 \times 120.2 \times (50/40.078) = 300$$
$$Mg = 2 \times 19 \times (50/24.31) = 78.16, \text{ and so on.}$$

Then run an ion balance:

$$\text{Cations} = \text{Anions or } 461 = 461$$

TABLE 5.3   Aquatic Ion Balance

Cation	Concentration (mg/l)	CaCO₃ Equivalent	Anion	Concentration (mg/l)	CaCO₃ Equivalent
Calcium	120.2	300			
Magnesium	19	78	Chloride	25	35.25
Manganese	0	0	Phosphate	52	55
Strontium	20	23	Bicarbonate	133	109.06
Iron		0	pH	7.5	0
Sodium	18	40	Nitrate	69	56
Potassium	15	20	Sulfate	198	216.32
Acidity					
SUM		**461**		**SUM**	**461**

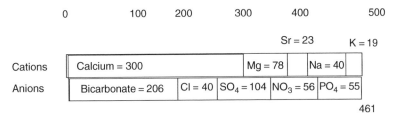

FIGURE 5.3   Plotted ion balance.

Next, create a diagram as shown in Figure 5.3 in the following orders:

Cations: calcium, magnesium, manganese, strontium, iron, sodium, potassium
Anions: bicarbonate, chloride, sulfate, nitrate, silicate, phosphate

$$\text{Total hardness} = 401 \text{ mg/l} \qquad \text{as } CaCO_3$$
$$\text{Bicarbonate alkalinity} = 206 \text{ mg/l} \qquad \text{as } CaCO_3$$
$$\text{Carbonate hardness} = \text{alkalinity} = 206 \text{ mg/l} \qquad \text{as } CaCO_3$$
$$\text{Noncarbonate hardness} = 401 - 206 \text{ mg/l} = 195 \text{ mg/l} \qquad \text{as } CaCO_3$$

CHEMICAL WATER SOFTENING

The process of removing hardness from water is important to many municipalities, especially in the Midwest and West where the groundwater has very high carbonate hardness. The information in brackets [ ] indicates the type of softening equations, for example, [LS1] is lime softening Equation 1.

Lime-Soda Softening (note underlined values indicate precipitates)

$$H_2CO_3 + Ca(OH)_2 = \underline{CaCO_3} + H_2O \qquad \text{[LS1]}$$
$$Ca^{2+} + 2HCO_3 + Ca(OH)_2 = \underline{2CaCO_3} + H_2O \qquad \text{[LS2]}$$
$$Ca^{2+} + 2Na_2CO_3 = \underline{CaCO_3} + 2Na \qquad \text{[LS3]}$$
$$HCO_3^- + Ca(OH)_2 = \underline{CaCO_3} + OH^- + H_2O \qquad \text{[LS4]}$$
$$Mg^{2+} + 2HCO_3 + 2Ca(OH)_2 = \underline{CaCO_3} + \underline{Mg(OH)_2} + 2H_2O \qquad \text{[LS5]}$$
$$Mg^{2+} + 2Ca(OH)_2 + Na_2CO_3 = \underline{CaCO_3} + \underline{Mg(OH)_2} + 2Na \qquad \text{[LS6]}$$

Note that the total hardness exceeds the total alkalinity. This is where lime-soda softening is applied.

Now develop the dosage equations for the reactions. Because we converted the dosages to equivalents of $CaCO_3$, the reactions are in milligram per litre per dose of $CaCO_3$. So looking back at the equations we get the following:

For the HCO_3 we need 103 mg/l of lime. For the balance of the calcium we need sodium bicarbonate in an amount of $(300 - 206) \times 2 = 188$ mg/l. For the manganese and strontium, we need 2 mg/l of lime and 1 mg/l of sodium carbonate per mg/l of each, so the lime dosage is $101 \times 2 = 202$ mg/l and the sodium carbonate is 101 mg/l.

The total lime dose is then $103 + 202$ mg/l $= 305$ mg/l
The total sodium carbonate is $188 + 101$ mg/l $= 289$ mg/l

Note that both of these compounds measured as CaCO₃ equivalents.

Conversions factors for lime and Na_2CO_3 are as follows:

1 mg/l lime $= 1.35$ mg/l of $CaCO_3$

1 mg/l sodium carbonate (solution) $= 0.94$ mg/l of $CaCO_3$

or 1 mg/l of sodium carbonate ($Na_2CO_3 \cdot 10\ H_2O$) (powder) is 0.35 mg/l of $CaCO_3$.

The residual materials in the water will be $Mg(OH)_2$ and $CaCO_3$ *at* their solubility products, plus sodium forms of the anions. The final pH of the water will be between 10 and 11. At a pH of about 10.3 the concentration of calcium and magnesium will be about 2.7 mg/l and 9.2 mg/l, respectively.

For removal of Noncarbonate hardness, NaOH can be substituted for Na_2CO_3. The problem with this process is the residual pH. At the high pH even with filtration, there is still a tendency for the residual carbonate to precipitate. So, the pH is reduced by adding either H_2SO_4 or CO_2 to a more normal range of pH between 8.5 and 9.5

Some useful equivalents are shown in Table 5.4.

EXCESS LIME PROCESS

The excess lime process is not a preferred process for removing Mg in water. It is primarily focused on the removal of carbonate alkalinity. One may use it where there is no significant noncarbonate hardness.

Given the following analysis:

pH $= 7.1$ Ca $= 180$ mg/l, Mg $= 60$ mg/l, Alk $= 260$ mg/l, Temp. $= 25°C$.

First calculate the carbonic acid concentration.

HCO_3 is all bicarbonate form because of pH. Alk $= 260$ mg/l.
now from *Standard Methods* (Method 4500 CO2)

$$K_1 = [H][HCO_3]/[H_2CO_3] = 10^{-6.36} \quad \text{or calculate from above}$$
$$K_2 = [H][CO_3]/[HCO_3] = 10^{-10.33} \quad \text{or calculate from above}$$

TABLE 5.4 Calcium Carbonate Equivalents

Substance	MW	Eq. Wt	Substance to CaCO$_3$ Equivalent	CaCO$_3$ Equivalent to Substance
Calcium Ca	40.1	20.05	2.50	0.40
Iron 2+ Fe^{2+}	55.8	27.9	1.79	0.56
Iron 3+ Fe^{3+}	55.8	18.6	2.69	0.37
Hydrogen H$^+$	1.01	1.01	50	0.02
Lead 2+ Pb^{2+}	207	103.5	0.48	2.07
Magnesium Mg^{2+}	24.3	12.2	4.12	0.24
Manganese Mn^{2+}	54.9	27.5	1.82	0.55
Nitrate NO$_3$$^-$	62.0	62.0	0.81	1.24
Sodium Na$^+$	23.0	23.0	2.18	0.46
Bicarbonate HCO$_3$$^-$	61.0	61.0	0.82	1.22
Chloride Cl$^-$	35.5	35.5	1.41	0.71
Sulfate SO$_4$$^{2-}$	96.1	48.05	1.04	0.96
Carbonic Acid H$_2$CO$_3$	62	31.0	1.61	0.62
Calcium Hydroxide Ca(OH)$_2$	74.1	37.1	1.35	0.74
Hydroxyl	17.0	17.0	2.94	0.34

Then all units are in CaCO$_3$

$$HCO_3 = [TALK - 5 \times 10^{(pH-10)}]/[1 + 0.94 \times 10^{(pH-10)}]$$

where TALK is Total Alk and CO$_3 = 0.94 \times [HCO_3] \times 10^{(pH-10)}$
and OH $= 5 \times 10^{(pH-10)}$ and Free CO$_2 = 2 \times [HCO_3] \times 10^{(6-pH)}$

Plugging the values in, we get
TALK $= 260$ mg/l, HCO$_3 = [260 - 0.5 \times 10^{(7-10)}]/[1 + 0.94 \times 10^{(7-10)} = 255.5/(1.00094) = 259.79$ mg/l
CO$_3 = 0.94 \times 259.79 \times 10^{-3} = 0.244$ mg/l, which is so small that it can be ignored and using the definition of K_1 above so $[H_2CO_3] = [H][HCO_3]/K_1 = 10^{-7} \times 0.425821/10^{-6.36} = 0.0955$ mol H$_2$CO$_3 = 153.85$ mg/l as CaCO$_3$ or
Find the carbonic acid fraction. Because alkalinity is defined in terms of CaCO$_3$ we need to back calculate to HCO$_3$$^- = 260$ mg/l   CaCO$_3 = X$ mg/l HCO$_3$$^-$ 50/eq. wt or $260 = X 50/61 = 317.2$ mg/l of HCO$_3$$^-$, which gives $317.2/61 = 5.2 \times 10^{-3}$ mol/l
Next calculate or estimate species fraction constants K_1 and K_2

$K_1 = 3.47 \times 10^{-7}$ $K_2 = 3.1 \times 10^{-11}$ and
$a_1 = [HCO_3]/Ct = 1/[10^{-7}/3.47 \times 10^{-7} + 1 + 3.1 \times 10^{-11}/10^{-7}] = 0.7761$,
and Ct $= 5.2 \times 10^{-3}/0.7761 = 6.7 \times 10^{-3}$ mol/l.
Then, since Ct $= H_2CO_3 + HCO_3 + CO_3$
Ct $= 155 \times 10^{-3}$ mol/l $= 155$ mg/l as CaCO$_3$

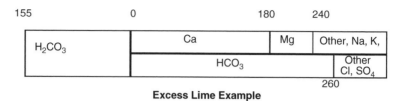

Excess Lime Example

FIGURE 5.4   Excess lime process treatment diagram.

The two results are quite similar, but the work getting there from *Standard Methods* is a bit easier. The challenge in the calculations is not to mix up the equivalents with the molar concentrations because there is a consistent set of conversions when you go back and forth to and from $CaCO_3$.

The next step is to draw the box showing the free acid and calculate the lime dose Fig. 5.4:

The lime dose $= H_2CO_3 + HCO_3 + Mg + 60\,mg/l = 155 + 260 + 60 + 60\,mg/l = 535\,mg/l$ as $CaCO_3$. The excess lime (60mg/l) is added to raise the pH above 11 to insure that the precipitation reactions go to completion.

The final pH of the water is about 11, and the hardness is between 30 and 50 mg/l of Ca plus about 10 mg/l of Mg hardness or a total of about 40–50 mg/l. This is relatively soft water.

A variant of this process is used for boiler waters. The hot lime softening process takes place near 100°C. The disadvantage is that the heat required is substantial, but the advantage is that the calcium and magnesium solubilities, which are noted for scale formation, are about one-third lower at the higher, temperatures, and because the water is hot it is less viscous, so the reactions take place faster, and the settling and separation are also a lot faster as well, and the process also hydrates silica as well.

Silica is objectionable in boiler waterfeed as well as in some municipal systems because of the scale formation potential. Silica is removed somewhat inefficiently along with magnesium hydroxide at the ratio of about 1 ppm of silica for 7–10 ppm of magnesium hydroxide precipitated. Silica control is often accomplished directly by ion exchange.

METALS REMOVAL BY PRECIPITATION

Dissolved metals are often removed by precipitation. Most commonly, the forms of the precipitants are hydroxides, carbonates, phosphates, and sulfides. Table 5.5 shows some important solubility products and can aid in the selection of metals removal by precipitation. Precipitation processes are, in general, not as good as ion exchange processes for metals removal, principally because the solubility product leaves some small quantity of the metal in solution, and in many wastewater applications, even with good precipitation, the formation of microflocs occurs, and these flocs are not removed efficiently, except through fine filtration.

HEAVY METALS

Heavy metals are important because they are often toxic, and they impede or interfere with the biological treatment process. Depending upon the metal and the species, all the reactions are pH dependent. When optimizing multiple metal removal in a waste species, it is often necessary to have a two-step process for pH removal. The chart (Fig. 5.5) indicates the difference in solubility of various metals for precipitation. This reference is relatively obscure but has been quoted widely and is available in the EPA Technology Transfer Seminar Publications–"Waste Treatment—Upgrading Metal Finishing Facilities to reduce Pollution—Volume 2".

Another interesting aspect of the diagram is that the solubility of a number of metals is pH dependent. For example, consider Cr and Zn. These metals are most often found in plating wastes and cooling towers. The optimum pH precipitation point of the metals is about 2 units apart. Zinc and Chromium may need separate pH tanks for treatment.

Also, when working with a carbonate precipitation, one must take extreme care that he or she understands the role of bicarbonate in the precipitation and the solubility of the precipitate with respect to pH.

CHROMIUM REDUCTION AND METALS PRECIPITATION

Hexavalent chromium, Cr^{6+}, has high aquatic toxicity and is a human and animal carcinogen. In the $6+$ valence, it is too soluble to be effectively removed by conventional precipitation. Therefore, it must be reduced either by reaction with Ferrous ions or by treatment with a sulfite such as Na_2S or with H_2S in gaseous form. Depending upon the chemical regimen used, the pH may have decreased to pH < 2.0 for the reaction to proceed. Then when the pH is raised the favored precipitation will be of Cr_2S_3, and the levels achieved after precipitation and filtration are extremely good, often as low as 0.007 mg/l–0.002 mg/l.

TABLE 5.5 Metals Solubility Products Solubility Constants for Inorganic Compounds @ 25°C

Compound	Formula	K_{sp}	Compound	Formula	K_{sp}	Compound	Formula	K_{sp}	Compound	Formula	K_{sp}
Aluminum hydroxide	$Al(OH)_3$	4.6×10^{-33}	Copper(I) iodide	CuI	1.1×10^{-12}	Magnesium arsenate	$Mg_3(AsO_4)_2$	2.1×10^{-20}	Silver iodide	AgI	8.3×10^{-17}
Aluminum phosphate	$AlPO_4$	6.3×10^{-19}	Copper(I) sulfide	Cu_2S	2.5×10^{-48}	Magnesium carbonate	$MgCO_3$	3.5×10^{-8}	Silver nitrite	$AgNO_2$	6.0×10^{-4}
Barium carbonate	$BaCO_3$	5.1×10^{-9}	Copper(II) arsenate	$Cu_3(AsO_4)_2$	7.6×10^{-36}	Magnesium fluoride	MgF_2	3.7×10^{-8}	Silver oxalate	$Ag_2C_2O_4$	3.6×10^{-11}
Barium chromate	$BaCrO_4$	2.2×10^{-10}	Copper(II) carbonate	$CuCO_3$	1.4×10^{-10}	Magnesium hydroxide	$Mg(OH)_2$	1.8×10^{-11}	Silver sulfate	Ag_2SO_4	1.4×10^{-5}
Barium fluoride	BaF_2	1.0×10^{-6}	Copper(II) chromate	$CuCrO_4$	3.6×10^{-6}	Magnesium oxalate	MgC_2O_4	7×10^{-7}	Silver sulfide	Ag_2S	6×10^{-51}
Barium hydroxide	$Ba(OH)_2$	5×10^{-3}	Copper(II) ferrocyanide	$Cu_2[Fe(CN)_6]$	1.3×10^{-16}	Magnesium phosphate	$Mg_3(PO_4)_2$	1×10^{-25}	Silver sulfite silver	$AgSO_3$	1.5×10^{-14}
Barium iodate	$Ba(IO_3)_2$	1.5×10^{-9}	Copper(II) hydroxide	$Cu(OH)_2$	2.2×10^{-20}	Manganese(II) carbonate	$MnCO_3$	1.8×10^{-11}	Silver thiocyanate silver	$AgSCN$	1.0×10^{-12}
Barium oxalate	BaC_2O_4	2.3×10^{-8}	Copper(II) sulfide	CuS	6×10^{-37}	Manganese(II) hydroxide	$Mn(OH)_2$	1.9×10^{-9}	Strontium carbonate	$SrCO_3$	1.1×10^{-10}
Barium sulfate	$BaSO_4$	1.1×10^{-10}	Copper(II) thiocyanate	$Cu(SCN)_2$	4.0×10^{-14}	Manganese(II) sulfide	MnS	2.5×10^{-13}	Strontium chromate	$SrCrO_4$	2.2×10^{-5}
Barium sulfite	$BaSO_3$	8×10^{-7}	Fluorapatite	$Ca_5(PO_4)_3F$	1.0×10^{-60}	Mercury(I) bromide	Hg_2Br_2	5.6×10^{-23}	Strontium fluoride	SrF_2	2.5×10^{-9}
Barium thiosulfate	BaS_2O_3	1.6×10^{-5}	Hydroxy-apatite	$Ca_5(PO_4)_3OH$	1.0×10^{-36}	Mercury(I) chloride	Hg_2Cl_2	5.0×10^{-13}	Strontium oxalate	SrC_2O_4	4×10^{-7}
Bismuthyl chloride	$BiOCl$	1.8×10^{-31}	Iron(II) carbonate	$FeCO_3$	3.2×10^{-11}	Mercury(I) chromate	Hg_2CrO_4	2.0×10^{-9}	Strontium sulfate	$SrSO_4$	3.2×10^{-7}
Bismuthyl hydroxide	$BiOOH$	4×10^{-10}	Iron(II) hydroxide	$Fe(OH)_2$	8.0×10^{-16}	Mercury(I) cyanide	$Hg_2(CN)_2$	5×10^{-40}	Strontium sulfite	$SrSO_3$	4×10^{-8}
Bismuth(III) sulfide	Bi_2S_3	1×10^{-97}	Iron(II) sulfide	FeS	6×10^{-19}	Mercury(I) iodide	HgI_2	4.5×10^{-29}	Thallium(I) bromate	$TlBrO_3$	1.7×10^{-4}
Cadmium carbonate	$CdCO_3$	5.2×10^{-12}	Iron(III) arsenate	$FeAsO_4$	5.7×10^{-21}	Mercury(I) sulfate	Hg_2SO_4	7.4×10^{-7}	Thallium(I) bromide	$TlBr$	3.4×10^{-6}
Cadmium hydroxide	$Cd(OH)_2$	2.5×10^{-14}	Iron(III) ferrocyanide	$Fe_4[Fe(CN)_6]_3$	3.3×10^{-41}	Mercury(I) sulfide	Hg_2S	1.0×10^{-47}	Thallium(I) chloride	$TlCl$	1.7×10^{-4}
Cadmium iodate	$Cd(IO_3)_2$	2.3×10^{-8}	iron(III) hydroxide	$Fe(OH)_3$	4×10^{-38}	Mercury(I) thiocyanate	$Hg_2(SCN)_2$	3.0×10^{-20}	Thallium(I) chromate	Tl_2CrO_4	9.8×10^{-15}

Name	Formula	K_{sp}
Cadmium sulfide	CdS	8.0×10^{-27}
Calcium carbonate	$CaCO_3$	3.8×10^{-9}
Calcium chromate	$CaCrO_4$	7.1×10^{-4}
Calcium fluoride	CaF_2	5.3×10^{-9}
Calcium hydroxide	$Ca(OH)_2$	5.5×10^{-6}
Calcium iodate	$Ca(IO_3)_2$	7.1×10^{-7}
Calcium oxalate hydrate	$CaC_2O_4 \cdot H_2O$	1.96×10^{-8}
Calcium hydrogen phosphate	$CaHPO_4$	1×10^{-7}
Calcium phosphate	$Ca_3(PO_4)_2$	1×10^{-26}
Calcium sulfate	$CaSO_4$	9.1×10^{-6}
Calcium sulfite	$CaSO_3$	6.8×10^{-8}
Chromium(II) hydroxide	$Cr(OH)_2$	2×10^{-16}
Chromium(III) hydroxide	$Cr(OH)_3$	6.3×10^{-31}
Cobalt(II) carbonate	$CoCO_3$	1.4×10^{-13}
Cobalt(III) hydroxide	$Co(OH)_3$	1.6×10^{-44}
Cobalt(II) sulfide	CoS	4.0×10^{-21}
Copper(I) chloride	$CuCl$	1.2×10^{-6}
Copper(I) cyanide	$CuCN$	3.2×10^{-20}
Iron(III) phosphate	$FePO_4$	1.3×10^{-22}
Lead(II) arsenate	$Pb_3(AsO_4)_2$	4.0×10^{-36}
Lead(II) azide	$Pb(N_3)_2$	2.5×10^{-9}
Lead(II) bromate	$Pb(BrO_3)_2$	7.9×10^{-6}
Lead(II) bromide	$PbBr_2$	4.0×10^{-5}
Lead(II) carbonate	$PbCO_3$	7.4×10^{-14}
Lead(II) chloride	$PbCl_2$	1.6×10^{-5}
Lead(II) chromate	$PbCrO_4$	2.8×10^{-13}
Lead(II) fluoride	PbF_2	2.7×10^{-8}
Lead(II) hydroxide	$Pb(OH)_2$	1.2×10^{-5}
Lead(II) iodate	$Pb(IO_3)_2$	2.6×10^{-13}
Lead(II) iodide	PbI_2	7.1×10^{-9}
Lead(II) sulfate	$PbSO_4$	1.6×10^{-8}
Lead(II) sulfide	PbS	3×10^{-29}
Lithium carbonate	Li_2CO_3	2.5×10^{-2}
Lithium fluoride	LiF	3.8×10^{-3}
Lithium phosphate	Li_3PO_4	3.2×10^{-9}
Magnesium ammonium phosphate	$MgNH_4PO_4$	2.5×10^{-13}
Mercury(II) sulfide	HgS	1.6×10^{-52}
Mercury(II) thiocyanate	$Hg(SCN)_2$	2.8×10^{-20}
Nickel(II) carbonate	$NiCO_3$	6.6×10^{-9}
Nickel(II) hydroxide	$Ni(OH)_2$	2.0×10^{-15}
Nickel(II) sulfide	NiS	3×10^{-19}
Scandium fluoride	ScF_3	4.2×10^{-18}
Scandium hydroxide	$Sc(OH)_3$	4.2×10^{-18}
Silver arsenate	Ag_3AsO_4	1.0×10^{-22}
Silver acetate	$AgC_2H_3O_2$	2.0×10^{-3}
Silver azide	AgN_3	2.0×10^{-8}
Silver benzoate	$AgC_7H_5O_2$	2.5×10^{-5}
Silver bromate	$AgBrO_3$	5.5×10^{-5}
Silver bromide	$AgBr$	5.3×10^{-13}
Silver carbonate	Ag_2CO_3	8.1×10^{-12}
Silver chloride	$AgCl$	1.8×10^{-10}
Silver chromate	Ag_2CrO_4	1.1×10^{-12}
Silver cyanide	$AgCN$	1.2×10^{-16}
Silver iodate	$AgIO_3$	3.0×10^{-8}
Thallium(I) iodate	$TlIO_3$	3.1×10^{-6}
Thallium(I) iodide	TlI	6.5×10^{-8}
Thallium(I) sulfide	Tl_2S	6×10^{-22}
Thallium(I) thiocyante	$TlSCN$	1.6×10^{-4}
Thallium(III) hydroxide	$Tl(OH)_3$	6.3×10^{-46}
Tin(II) hydroxide	$Sn(OH)_2$	1.4×10^{-28}
Tin(II) sulfide	SnS	1×10^{-26}
Zinc carbonate	$ZnCO_3$	1.4×10^{-11}
Zinc cyanide	$Zn(CN)_2$	3×10^{-16}
Zinc hydroxide	$Zn(OH)_2$	1.2×10^{-17}
Zinc iodate	$Zn(IO_3)_2$	3.9×10^{-6}
Zinc oxalate	ZnC_2O_4	2.7×10^{-8}
Zinc phosphate	$Zn_3(PO_4)_2$	9.0×10^{-33}
Zinc sulfide	ZnS	2×10^{-25}

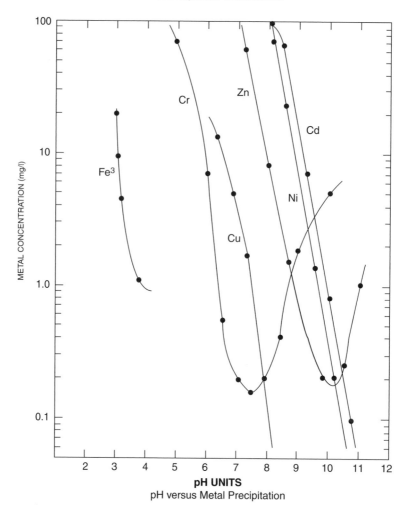

FIGURE 5.5 Metals solubility at various pH. *Source*: USEPA Electroplating and Metal Finishing Source Control Manual, citing: R. Weiner, Die Abswaser der Galvanotechnik und Metallindustrie, 4th Edition Eugen G. Leuze Verlag, 1973.

Sulfite precipitation of metals is excellent but often creates a toxic waste disposal problem. More often than not, the metals are hazardous wastes by classification and can be disposed of only after solidification and/or reclamation.

Cadmium is especially difficult to reclaim and even worse to dispose of. It is moderately easy to precipitate either as a hydroxide or as a sulfide. One of the principal sources of cadmium is from plating solutions where it is used to enhance the flexibility and corrosion resistance of the plating being applied.

Because of its toxicity and relatively low melting point, cadmium is moderately dangerous to smelt for recovery. The popularity of Ni–Cd batteries has made this problem even more difficult. No one wants Cd but everyone takes the Ni. Currently, in the United States there are no Cd recovery facilities. The closest cadmium reclamation facilities are in England and Japan, and Cd requires an export license from the EPA (Table 5.5).

SILICATES IN TREATMENT SYSTEMS

Silicates do belong to a treatment regimen but not necessarily for the raw water supply or for boiler water. Silicates are often useful in assisting coagulation in a waste treatment system. In developing a chemical regimen, you will also want to look at the chemistry of silicates as part of the precipitation and/or coagulation system. Silica forms a number of insoluble precipitates with a wide variety of metals. The addition of a few parts per million of silicates to a precipitation system not only enhances the precipitation but also provides nucleation for precipitation as well as a more dense and easily settled sludge. The problem is getting the silicates into solution.

Fortunately, that is relatively simple. One can make up a silica sol solution with almost any compound. Take a dilute solution of water glass (sodium silicate liquid) about 5–30 g/l and back titrate it to near neutral (pH < 8) with an acid, and then finish the titration to about pH 7.3 with an amphoteric metal such as aluminum. If you use chlorine as a titrant, you will have a chlorine compound as a sol, and if you use alum, you will have an oil and grease breaking sol, which actively removes calcium and phosphorous and also makes a good nucleating agent for flocs.

Silica sols were widely used in the older technology but abandoned when high charge and high molecular weight polymers were developed. The sols still have been found to be cheaper and work better in some instances. The Philadelphia Quartz company, www.pq.com, has a useful brochure on the subject, which is included in a Disk and is provided as a supplement this book.

NITROGEN

The nitrogen series is important. When ammonia oxidizes to nitrate, it requires substantial amounts of oxygen. The first oxidation is to nitrite by nitrosomonas bacteria. Nitrite is toxic and is often used as a food preservative. The second group of bacteria take the nitrite and oxidize it to nitrate. The rate constant for conversion from nitrite to nitrate is about three times faster than the conversion rate between ammonia and nitrite. As a consequence, nitrite concentrations in a viable bacterial population are seldom above 0.1 mg/l, and also because one of the measurement techniques for nitrate also converts the nitrate to nitrate, the results are often reported only as nitrate plus nitrite, or equally often the nitrite is ignored and the results are reported only as nitrate.

The equations are as follows:

$$2\,NH_3 + 3.5\,O_2 = 3\,H_2O + 2\,NO_2$$

$$2\,NO_2 + O_2 = 2\,NO_3$$

$$2\,NH_3 + 4.5\,O_2 = 2\,NO_3 + 3\,H_2O$$

Nitrate is also a key oxygen source during marginal oxygen conditions in the stream or in the wastewater treatment plant. Under reduced oxygen conditions, facultative and other aerobic bacteria will strip the oxygen off nitrate to continue their preferred aerobic processes. In many new wastewater treatment systems, the toxicity of ammonia is of concern to the aquatic environment. The excess nitrogen (not needed for biological growth) is removed by operating a wastewater treatment under anoxic, nitrogen-reducing conditions. These conditions will take the excess nitrogen back all the way to nitrogen gas where it is re-released to the atmosphere.

SULFUR

Sulfur is also a very important compound in the aquatic environment, but it shows up principally as the sulfate or HSO_3 form as a compound or as the salt of sulfuric acid. When anaerobic conditions are present, the sulfate is reduced to H_2S, which is both toxic and volatile. The reduction of sulfur takes place after the nitrate has disappeared.

PHOSPHOROUS

Phosphate is important because it has been found to be a key ingredient in creating algae blooms downstream of treatment works. In a later chapter we will discuss removal of phosphorous by chemical precipitation and by biological means. This latter process is known as luxuriant uptake of phosphorous. The chemical means of precipitation is often more reliable. Phosphorous is removed by precipitation with iron, lime, aluminum. The reactions are straightforward and well documented and will be discussed when we get to the chapter on phosphorous removal treatment.

6

ELEMENTS OF BIOLOGICAL TREATMENT

INTRODUCTION

The purpose of this chapter is to provide you, the reader, with a brief overview of the most important methods of wastewater treatment plant design currently in use. The first method is the older rationale method, which is embodied in the Ten States Standards and other design codes.[1] The second

---

[1]Ten States Standards are available from Health Education Services in Albany, NY at http://www.hes.org. The latest edition of the standards is the 2004 edition.

---

is based upon the Monod equation. However, before we get started, we need to define some elemental terms important to the industry.

BOD AND COD SOLIDS

There is a difference between BOD and COD, as outlined briefly in previous chapters. BOD is based on dissolved oxygen reduction by *acclimated organisms*[2] in consuming (oxidizing) organic carbon in the wastewater. This is critical to our definitions. COD is based upon chemical oxidation of all organic carbons using an acid dichromate oxidation or in some countries a permanganate oxidation of organics. The dichromate method is the standard in the United States, but when comparing data from non-U.S. sources, and even older data, you should check the methodology used for performing the COD test, because the permanganate oxidation will give consistently lower results. Certain inorganic substances, such as sulfides, sulfites, thiosulfates, nitrites, and ferrous iron are oxidized by dichromate, creating an inorganic COD, which is misleading when estimating the organic content of the wastewater and can yield high results. The standard estimate of the ratio between BOD and COD for domestic sewage and plant sanitary wastewater is that the BOD is about 0.64–0.68 of the COD for the same sample.

In looking at the comparison between BOD and COD and the issues surrounding them, the following rules generally apply:

1. The COD is always higher than the BOD, and the COD will always oxidize things that the BOD cannot or will not measure.
2. In any given wastewater, there is a likelihood that a small portion of the COD will be refractory caused by oxidation of thiosulfates, sulfides, and other compounds. This refractory COD cannot be efficiently or effectively removed from wastewater.
3. This is not true for BOD. Several eminent practitioners have proposed the idea of refractory BOD. If one can degrade materials in a BOD bottle, they can be degraded in a wastewater treatment plant. Refractory BOD does not really exist. Instead, it represents that small fraction of BOD that is uneconomical or impractical to treat in a wastewater treatment plant.

---

[2]The use of acclimatized organisms is one of the major weaknesses in using BOD as a regulatory measurement parameter. When industrial effluents are sampled, and where those effluents contain biologically resistant or unusual compounds, it is virtually impossible to obtain an accurate BOD measurement because the "seed" organisms have not developed extracellular enzymes that will permit them to degrade the compounds in the industrial effluent. As a consequence, the BOD measurements will be disproportionately low and unrepresentative of the true strength of the waste system.

TABLE 6.1   Relationships between BOD, COD, and ThOD^a

Compound	Theoretical Oxygen Demand	Measured COD	Measured BOD	BOD/COD	BOD/ThOd
Ethanol	2.080	2.110	1.580	0.749	0.760
Ethylene glycol	1.260	1.210	0.360	0.298	0.286
Maleic acid	0.830	0.800	0.640	0.800	0.771
MEK	2.440	2.200	1.810	0.823	0.742
Methanol	1.500	1.050	1.120	1.067	0.747
O-Cresol	2.520	2.380	1.750	0.735	0.694
Toluene	3.130	1.410	0.860	0.610	0.275

^{a}All values in mg/mg

4. The general relationship between BOD and COD for sewage and most human wastes is about 1 unit of BOD $\simeq$ 0.64–0.68 units of COD. The relationship is not consistent because of the variable quantity of solids and soluble carbon in sewage. The common interferences for COD, which cause it to be higher than BOD include sulfides, sulfites, thiosulfates, and chlorides.

5. The BOD test must be inhibited to prevent oxidation of ammonia. If the inhibitor is not added, the BOD will be between 10% and 40% higher than can be accounted for by carbonaceous oxidation.

The COD is closer but *not equal* to the theoretical oxygen demand or ThOD. See Table 6.1 for a presentation of a number of chemicals.

As will be discussed in the following sections, wastewater models, especially dynamic models, define all parameters in terms of COD. Moreover, the wastewater modeling field uses a slightly different definition based on filtration through a 0.2 μm filter, rather than 0.45 μm filter used to measure suspended solids. On this basis, the modelers separate suspended material and dissolved material.

SUSPENDED SOLIDS

All waste streams have some suspended solids. Domestic Sewage, depending upon strength will run from 150 to 250 mg/l. Petrochemical wastes can have TSS in excess of 400 mg/l. Some waste streams, including paper plants, food wastes, and some petrochemical processes, have TSS loads in excess of 1000 mg/l.

The solids generally have a biodegradable component and may have active biomass, again depending upon the process. Often the solids represent

TABLE 6.2   Typical Contents of Sewage

Fraction	Total Solids (%)	Volatile Solids (%)	Organic Nitrogen (%)	COD (%)
Settlable Solids	18	28	23	34
Supracolloidal	13	22	27	27
Colloidal	62	37	42	25
Totals in Mg/L	400	200	15	250

Source: Eckenfelder and Ford: Water Pollution Control procedures for process design

between 30% and 60% of the BOD, but for industrial wastes, that figure is highly variable. The strength of the wastes depends directly upon the amount of water used within the plant.

The suspended solids are a collection of organic and inorganic materials of various sizes and density. The size and density ranges are from 3–5 mm to 0.001 mm, and from 0.8. to 2.65 gm/cm^3 and higher. It should be noted that the latter value for density is primarily owing to sand and clay; in industrial wastes, the higher density particles may include metal scraps, (machining) bolts, screws, nails, and so on. The larger particles tend to be cigarettes, insects, various types of floating solids, and even food particles. The smaller particles are often bacteria, which tend to be indistinguishable or invisible in the water.

In domestic sewage, one often finds the values as shown in Table 6.2.

In industrial wastes, these relations do not hold because of complex chemicals and the lack of substantial human fecal matter discharges.

BIOLOGICAL GROWTH EQUATION

The following are general principles that you should observe in dealing with a biological treatment system.

Before we get into the biological growth equation, we need to look at the balance of the waste stream. For ideal biological growth, the waste should be balanced.

1. The carbon:nitrogen:phosphorus (C:N:P) ratio of sewage is often ideal. Look closely at the C:N:P ratio of the industrial wastes, because it should be between 100:20:1 and 100:5:1 for ideal biological growth.
2. If the C:N:P ratio of the waste is strong in one direction or the other, poor treatment will result. This is especially true if the waste is too strong in carbon.

3. The waste should also be neither too weak nor too strong; although too weak is acceptable, it is difficult to treat. This is extremely hard to define. BOD is best treated in the range of 60–500. The range of the upper limit is primarily a limit on aeration ability of the system. Wastes in excess of 500 mg/l BODs have been treated very successfully if sufficient dilution is practiced in the treatment process, or if anaerobic processes are used for pretreatment. One aeration process uses high purity oxygen in the aeration system, rather than air, because the high purity oxygen has a greater transfer efficiency and is more suitable for treating higher strength wastes.

4. Biological treatment is easily effective in removing 95–98% of the BOD, but if you need to go beyond that limit, additional measures may have to be implemented.

5. You cannot get all the BOD removed in a biological treatment system without extremely large tankage and that may be uneconomical.

6. You will not get all the COD removed from the waste treatment system for the reasons cited above. But also because a part of the COD is nonbiodegradable. That is the only case for refractory COD.

7. Biological treatment systems do not handle shock loads well. Pretreatment or equalization may be necessary if the variation in strength of the waste is more than about 150% or if that waste at its peak concentration is in excess of 1000 mg/l BOD.

8. Biological systems do not like extreme variations in hydraulic loads either. Diurnal variations of greater than about 250% may be a problem primarily because they will create biomass loss in the clarifiers.

9. Toxic and biologically resistant materials will require special consideration and may require pretreatment before they are discharged into the wastewater treatment plant.

10. Oils and solids do not belong to a wastewater treatment system because they interfere with the treatment. Pretreat these wastes to remove inert solids, oils, and excessive biological solids of more than 200 mg/l–300 mg/l.

11. The capacity of the aeration system you will use is finite with regard to oxygen transfer. The capacity of the waste to use oxygen is unlimited. Consider this in the design.

12. The growth rate of biological organisms is highly temperature dependent. A 10°C reduction in water temperature can cut the biological reaction rates in half. Wintertime conditions where the

water is cold will virtually stop all nitrification and will slow all of the other biological processes as well. If you are dealing with cold climates you may want to consider covering the wastewater treatment tanks to help preserve the little heat left in the system, and if the system is small enough, you may want to consider heating the wastewater.

BIOLOGICAL GROWTH & THE MONOD EQUATION

Biological Growth can be described according to the Monod equation:

$$\mu = (\lambda S)/(K_s + S)$$

where μ = specific growth rate coefficient; λ = maximum growth rate coefficient, which occurs at 0.5 μ_{max}; S = concentration of limiting nutrient, that is, BOD, COD, TOC, and so forth; K_s = Monod coefficient. This is also called the half-saturation coefficient because it corresponds to the concentration at which μ is half of its maximum. This can be seen from the Monod equation by setting S equals to K_s. K_s ocurs at λ (0.5μ_{max}).

The curve in Figure 6.1 is a plot of specific growth rate coefficient versus concentration of growth-limiting substrate when there is no inhibition.

Organics + Bacteria + Nutrients + Oxygen
    $\rightarrow$ New Bacteria + CO_2 + H_2O + Residual Organics + Inorganics

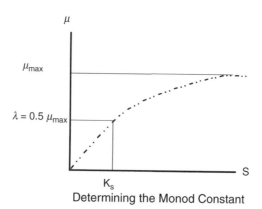

Determining the Monod Constant

FIGURE 6.1   Determining the Monod growth rate coefficient.

Supplemental Discussion of Biological Treatment

The Monod equation is:

$$\mu = \lambda S/(K_s + S)$$

Using this we will consider the development of the ideas behind the reaction rates and the meanings of a few terms relating to all biological decay. Much of the source of what follows is from *Activated Sludge Systems* by Orhon and Artan, published by Technomic Press, ISBN Number 1-56676-101-8. The book is about 600 pages of everything one would like to know about kinetic reactor modeling and activated sludge but were afraid to ask.

Another way of looking at the overall stoichiometric rate is

$$R_g = R_e + R_s$$

where R_g is the overall growth reaction, R_e is the energy reaction, and R_s is the biosynthesis reaction. The subcomponents are as follows: one reaction for biomass R_c, one for electron donor R_d, and one for the electron acceptor R_a. The reactions can be further resolved into half reactions with the following:

$$R_s = Y(R_d - R_c)$$

where R_s is the biosynthesis reaction as shown above, and Y is the yield of the reaction. From here we can go into stoichiometry and balanced chemical reactions and half reactions for various chemicals. We are NOT going to do that because it is in a way giving more information than you need.

If the overall reaction rate for organic growth is

$$R_o = R_e + R_s + R_{de}$$

where R_e is the energy reaction shown above, and R_s is the synthesis reaction, and R_{de} is the Decay reaction, $R_{de} = R_c - R_a$.

Going back to the Monod equation:

The change in substrate with time is

$$dX/dt = \mu(SX)/(K_s + S)$$

and the specific maximum substrate removal rate is defined as $k_{max} = \mu/Y$

As the substrate value gets very large, the Monod curve tends to flatten out at the top and becomes a straight line, which is at μ_{max}. This is true where $S \gg K_s$. At that point, the growth is essentially limited by the ability of the bacterial population to transfer food through their cell walls and reproduce. This is not the case for most activated sludge systems, because other limiting factors such as oxygen transfer and availability of other nutrients tend to govern and have caused heavily loaded substrate systems to fail and go anaerobic.

Where there are two wastewaters generated on an alternating basis, each is likely to have its own specific growth rate and kinetics. For treatment purposes, the substrate with the slower growth rate will govern the design. For blended wastes, look at the blend and apply the slowest growth rate accordingly.

Microbial Decay   The loss of cellular mass is microbial decay where cell death occurs. Essentially it is the degradation of endogeneous mass for the generation of maintenance energy. It is the second part of a set of sequential processes where all exogeneous substrate is first used for synthesis of cell material and later decays as the cell ages and substrate concentrations decrease. The decay process is measured by changes in particulate matter in the system, by the change in the mixed liquor volatile suspended solids, or *MLVSS* or simply *VSS*. This is shown by the rate equation

$$\mathrm{d}X/\mathrm{d}t = -k_\mathrm{d}X$$

where k_d is the endogenous decay coefficient $[\mathrm{T}^{-1}]$; and X is the volatile suspended solids concentration.

VSS measures many things and not just the specific decay, and it is a very broad parameter for estimating kinetic coefficients, with an accuracy of about $\pm 20\%$ or less.[3] The growth rate constants can be significantly different for similar wastewaters, even for domestic wastes. Part of the reason for this difference is the internal differences in composition of organic matter and dissolved materials that may not show up without more extensive testing.

Industrial wastes data are also shown in Table 6.4. Typical constants are given for a variety of chemicals. The consistency is slightly greater for industrial wastewater, but not much better. The data are limited and the fact that the values are much more consistent may have as much to do with the idea that the wastes are predominantly one product as the fact that there is a much smaller database to work with. Sometimes, you may only have one value (see Tables 6.3 and 6.4), and may have to make an educated guess.

---

[3]The accuracy of the suspended solids test varies inversely with concentration from 33% at 5 mg/l TSS to 0.76% at 1707 mg/l TSS. There are no published accuracy data for the VSS test.

TABLE 6.3   Various Kinetic Constraints for Domestic Wastewater

Basis for Constants	μ (day^{-1})	K_s (mg/l)	K_d (day^{-1})	Y
BOD$_5$	0.6	12–80	0.01–0.14	0.38–0.68
BOD$_5$	6	100	0.048–0.055	0.5–0.67
BOD$_5$	1.43–13.2	25–120	0.04–0.075	0.42–0.75
COD	1.70	43–223	0.016–0.068	0.31–0.35
COD	3.75	22	0.07	0.67
COD	3.20–3.75	22–60	0.07–0.09	0.4–0.67

The value of Y is calculated from VSS data (Source: Modeling of Activated Sludge Systems, op. cit.)

Effect of Temperature on Rate of Reactions   Temperature corrections for the rate of reaction have been simplified from the Arrhenius equation to a much more simple form, and depending upon the model you are using to look at the wastewater, you will have to correct for temperature. The standard is at 20°C. The correction for temperature is

$$r_t = r_{20}\theta^{(t-20)}$$

where r_t is the reaction rate at temperature t and θ is the temperature coefficient. In the ranges of $t = 10°C–40°C$, θ has a value of between 1.0 and 1.10, with a common value of 1.04.

pH Effects   Low pH can stop a biochemical reaction or reduce its rate to almost zero. Orhon and Artan give the following formula for the effects of pH on growth rate. The equation is

$$\mu = \mu' K_i/(K_i + H^+)$$

where μ' is the original uptake rate; K_i = disassociation rate constant for the second reaction constant for the substrate, that is, E = enzyme, S = substrate, and the reactions are as follows:
   $E + S \leftrightarrow ES$ and $ES + H^+ \leftrightarrow ESH^+$ and $ESH^+ + H^+ \leftrightarrow ESH_2^{2+}$ where K_i is the disassociation constant for ESH_2^{2+}.

TABLE 6.4   Kinetic Constraints for Industrial Wastes

Industry	μ (day^{-1})	K_s (mg/l)	Y	K	Basis for Constants
Textile	0.1–6.96	86–95	0.52–0.73	0.013–0.12	BOD$_5$
Poultry		500	1.32	0.72	BOD$_5$
Soybean	12	355	0.74	0.144	BOD$_5$
Meat processing	0.57–1.09	150–362	0.34–0.42	0.03–1.0	COD
Edible oil	0.36	350	0.28	0.075	BOD$_5$
Skim milk	2.45–2.9	100–110	0.48–0.50	0.45	BOD & COD

Y is calculated on a VSS basis (Source: Modeling of Activated Sludge Systems, op. cit.)

PRINCIPLES OF BIOLOGICAL TREATMENT SYSTEMS

Consider the schematic drawing of the following system shown in Figure 6.2. This is a classical representation of an activated sludge system. We will examine the mass balance and look at the equations for biological growth.

The first task is to set out the terms, which we will use in analyzing the flows in the system.

Q = Volumetric influent rate (volume/time)

Q_w = Waste sludge volumetric flow rate (volume/time)

Q_3 = Effluent flow rate (volume/time)

Q_r = Recycle flow rate (volume/time)

X_1 = Microorganism influent concentration (mass/volume influent)

X_2 = Aeration basin microorganism concentration (mass/volume)

X_3 = Secondary effluent microorganism concentration (mass/volume)

X_r = Recycle and wasted solids concentration

V_2 = Aeration basin volume

r_{BH} = Reaction rate for solids also may be written as dX/dt = rate of change of microorganisms concentration in aeration basin (mass/volume time)

r_s = Reaction rate for substrate.

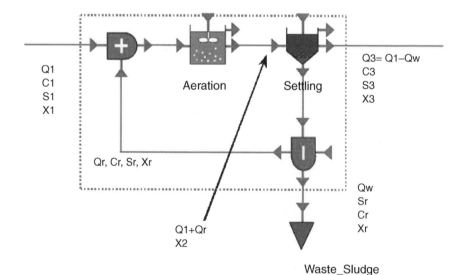

FIGURE 6.2   Basic schematic of activated sludge system.

Rate of bacterial growth $r_{BH} = \mu X$, where X is the microorganism concentration in mass/volume and μ is the specific growth rate per unit of time

$$\text{Cell yield coefficient} = Y_{obs} = \frac{-r_g}{r_{su}}$$

where $Y_{obs} = $ observed yield coefficient and $r_s = $ substrate utilization rate; $r_{BH} = $ cell growth rate

and

$$r_{BH} = -Y_{max} r_s - bX$$

where Y_{max} is equal to λ and b is the specific maintenance rate, endogenous or decay coefficient in units of time.

This gives us a sample of a solution for a steady state system.

Now, when we look at a biological treatment system, we will consider a simple system comprises a reactor or aeration tank and a clarifier or solids removal device, as shown in Figure 6.2.

Running a balance around the system we get the following:

$$Q_1 X_1 + V X_2 r_2 = Q_3 X_3 + Q_w X_r$$

if X_1 is relatively small with respect to X_2 and we assume steady state operations, then the equation becomes

$$\mu = r_2 = \frac{Q_3 X_3 + Q_w X_r}{V_2 X_2}$$

For a bioreactor, mean cell residence time = sludge age = θ_c = solids mass/ change in solids mass = $X/(\delta X/\delta t)$ or

$$\theta_c = 1/\mu = 1/r_2 \frac{V_2 X_2}{Q_3 X_3 - Q_w X_r}$$

One measure of activated sludge systems is the mean cell residence time or sludge age. The different types of systems and much of U.S. terminology are involved with sludge age.

Again, at steady state conditions and making a substitution from above we get

$$Y_{obs} = \frac{\theta X}{\theta_c (S_o - S)} \quad \text{and} \quad \frac{Y_{max}}{1 + b\theta_c} = \frac{\theta X}{\theta_c (S_o - S)}$$

$$\text{Specific utilization rate} = U = \frac{S_o - S}{\theta X}$$

With one other critical substitution of Efficiency

$$E = \frac{S_o - S}{S_o} \times 100$$

we get $U = F/ME \times 10^{-2}$

where F/M is the food to microorganism ratio or

$$F/M = S_o/\theta X$$

The F/M ratio is one of the key parameters in designing an aerobic treatment system by conventional means in the United States. This is also called loading rate.

ACTIVATED SLUDGE & ITS VARATIONS

The parameters that hold for aerobic treatment systems are shown in Table 6.5.

TABLE 6.5   Wastewater Treatment Plant Definitions by Loading Rate

Aerobic Wastewater Treatment Characterizations		
Process	Loading Rate[*]	Removal Efficiency
Extended air	0.05–0.20	85%–95%
Conventional activated sludge	0.2–0.5	90%–95%
Contact stabilization	0.2–0.5	85%–90%
High rate stablization	0.5–5	60%–85%

[*]The definition is Q_{So}/S_{V1}

Another often useful measure of the aeration system is by defining the X_{BH} in the aeration tank and the volumetric holding time or θ. This gives us the classification scheme for types of plants and their configurations as shown in Figure 6.3.

Typical design parameters for activated sludge process modifications are shown in Table 6.6.

TABLE 6.6   Typical Design Parameters for Activated Sludge Process Modifications

Modification	Process Loading Range	MLSS, mg/l	Aeration Time (h)	R/Q, Percent
Complete mix	Conventional low rate	3000–6000	3–5	25–100
Plug flow	Conventional low rate	1500–3000	4–8	25–50
Contact stabilization	Conventional rate	1000–3000[a]	0.5–1.0[a]	25–100
		4000–10,000[b]	3–6	
Step feed	Conventional rate	2000–3500	3–5	25–75
Extended aeration	Low rate	3000–6000	18–36	75–150
Oxidation ditch	Low rate	3000–5000	18–36	75–150
High purity oxygen	High conventional rate	3000–5000	1–3	25–50

[a]Contact tank.
[b]Stabilization tank R/Q is equivalent to Q_r/Q_1

Types of Activated Sludge Processes

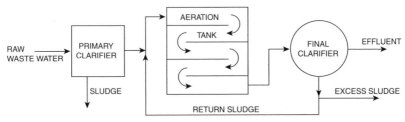

CONVENTIONAL ACTIVATED SLUDGE PLANT

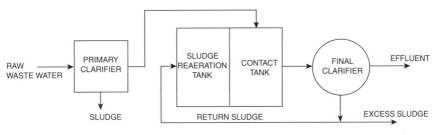

COMPLETE MIX PLANT

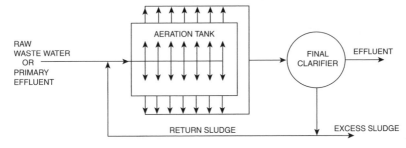

CONTACT STABILIZATION PLANT

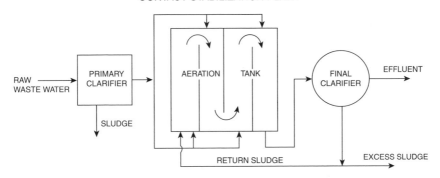

STEP AERATION PLANT

FIGURE 6.3 Basic wastewater plant definitions.

Substrate Removal Definitions:

Water Balance is $Q_1 = Q_3 + Q_w$
Running a mass balance around the system we get

$$Q_1 C_1 - r_{vs} \times V_2 = Q_3 \times C_3 + Q_w \times C_r$$

This time, the term r_{vs} is both negative and defined in terms of the volume of the substrate in the tank. We could just as easily have defined the term as $r_s S$.

This is a simple way of saying that the bacterial growth removes substrate from the tank. Note that in the above equation no specific definitions are implied, so C can be NO_2, NH_3, COD, or anything else. However, one must define the unit of volume as well as the reaction rate. This means that the reaction rate can be r_{vs} or r_{xs} together with the unit of the volume V_2 and for r_{sx} the activated sludge concentration X_2.

The units must be internally consistent. That is to say that the activated sludge concentration X_2 can be measured in kg of SS/M^3, kg of VSS/M^3, or kg of COD/M^3, but the units must be consistent in the numerator and denominator.

At steady state the materials must be all hydrolyzed before they can be accessed and consumed by the bacteria. So if you have some substrate such as BOD or COD it cannot be used until it is solubilized. That means that the basic balance will look like the following when we consider the same mass balance as in Figure 6.3, only we have now added the growth and hydrolysis terms to the equations:

If we look only at the system boundaries shown in the box, in Figure 6.2

$$\text{Input} + \text{Hydrolysis} + \text{Growth} = \text{Effluent} + \text{Sludge wasting}$$

$$Q_1 S_{s1} + K_h X_{S2} V_2 + (-(1/Y))\mu[S_{s2}/(S_{sw} + K_s)][S_{O_2}/(K_{sO_2} + S_{O_2}))X_{BH} V_2$$
$$= Q_3 S_3 + Q_w S_r$$

where X_{BH} is the heterotrophic biomass, S_{O_2} is the oxygen concentration, as noted earlier. Note that we are taking a balance around the system and not just around the aeration tank. The growth term removes the substrate.

When we look at specific variables in the activated sludge process, we can begin to write equations for mass balances of specific parameters. Fortunately, a number of researchers have already examined the activated sludge process and prepared a summary of critical terms and constants. These are codified in Activated Sludge Model No. 1.[4]

---

[4]Activated sludge model no. 1 (ASM1) is published by the International Water Association, Alliance House, 12 Caxton St., London, SWH 05Q, UK.

TABLE 6.7 Formulation for "Parameter Sensitive Switches" in Activated Sludge Kinetics

Parameter	Variable	Sample
DO	DO, K_{DO}	$= DO/(K_{DO} + DO)$
Ammonia	NH_3, K_{NH_3}	$= NH_3/(K_{NH_3} + NH_3)$
Nitrate	NO_3, K_{NO_3}	$= NO_3/(K_{NO_3} + NO_3)$
Alkalinity and pH	ALK, K_{ALK}	
	K_{pH}, 1	$= K_{pH}/(K_{pH} + 1)$, where K_{pH} is the pH Constant, and $1 = 10\exp(\text{optimum pH-pH}) - 1$

Because oxygen concentration is critical for aerobic substrate removal, the Monod term has been added for dissolved oxygen. We can also add other Monod equation terms to the equations to compensate other parameters as well. Monod equation is $S/(S + K)$. A few of those are shown below and several of them act as "switches", because when S goes below a certain specific value the term tends toward zero and the entire multiplier falls out of the balance equation.

Some of those terms are shown below in the following formula:

$$X/(K + X)$$

where X is the parameter and K is the half saturation constant. The value and a sample formulation is shown in Table 6.7:

Any number of the switches can be included on many of the models to account for optimum performance.

The equations above are often expressed in the form of a Petersen Matrix for the ease of writing. The table is read for rate equations both down and across, and a part of the matrix is presented in Table 6.8.

TABLE 6.8 The Petersen Matrix for Activated Sludge Equations

Component	S_s	X_s	X_l	X_{BH}	S_{O_2}	Reaction Rate r_v
Process						
Aerobic hetrotrophic growth	$-(1/Y)$			1	$(1-Y)/Y$	$\mu(S_s/(S_s + K_s))\,(S_{O_2}/(K_{S_{O_2}}+S_{O_2}))\,X_{BH}$
Hetrotrophic decay		$1-f_{XB}$	f_{XB}	-1		$b_H\,X_{BH}$
Hydroloysis	1	-1				$K_h\,X_S$
Units				kg COD/m³		
					Oxygen	
				Hetrotrophic biomass		
			Inert suspended organic matter			
		Slowly degradable organic matter				
	Easily degraded organic matter					

This is the start of the formulation of the equations for most of the activated sludge models developed by the International Water Association (IWA).

Trickling Filters and Variations

A second and older type of treatment still in use is the trickling filter, or in some instances it is also used as a roughing filter for pretreatment. The filter itself is not really a filter but an attached growth platform for microorganisms. It generally consists of large rocks, or plastic media with a large surface area, and the waste is sprayed, dumped, or poured over the filter in an intermittent fashion. The intermittent nature of the flow is to permit the organisms to breathe.

The filter bed does not really filter at all, and can be anything from engineered plastic media to crushed rock. The entire purpose of the filter bed is to serve as a support platform for the bacteria that grow out of it and to provide them with a void space so that the surface of the liquid has an opportunity to contact the atmosphere, where it can transfer oxygen into the liquid in support of the bacteria.

Figure 6.4 shows some typical diagrams of trickling filter systems in current use.

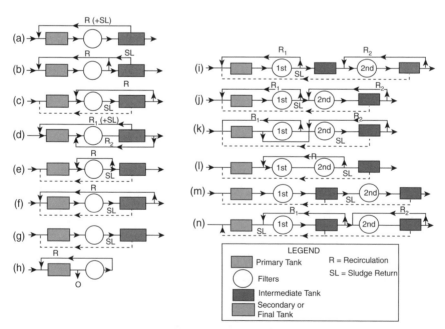

Flow diagrams of single and two-stage trickling filter plants.

FIGURE 6.4   Typical configurations for trickling filters. *Source*: WEF MOP/8 Waste Water Treatment Plant Design Manual of Practice number 8.

TABLE 6.9   Properties of Trickling Filter Media

Media/Packing	Nominal Size (in.)	Units per ft^3	Unit Weight per ft^3	Dry Specific Surface Area ft^2/ft^3	Void Space (%)
Plastic media	20 × 48	2–3	2–6	25–35	94–97
Redwood Media	47.5 × 47.5 × 1.8		10.3	14	80+
Granite and stone	1–3		90	19	65
Blast furnace slag	2–3	51	68	20	49

In the trickling filter, it is important to have a medium that has a large surface area with respect to the volume of the media. Table 6.9 illustrates some typical properties of trickling filter media.

The efficiency of trickling filters is calculated in a number of ways. The most readily understandable is the Eckenfelder formula.

$$S_e/S_o = e^{-X}/\{(1 + N) - N_e^{-X}\}$$

where S_e and S_o are as defined above, N = hydraulic recycle ratio, $X = KD^m/Q^n$, K = specific surface area (ft^2/ft^3) × removal rate constant, D = depth, Q = hydraulic loading, and m and n = determined media constants. For most applications $n = 1$.

Most trickling filters are extremely temperature sensitive, because they rely on direct contact with air and their performance follows the power law about biological activity and temperature, that is, the activity doubles or halves for each 10°C change in temperature.

Author's Comment: The trickling filter is still in use, but inherent limitations and the great costs associated with its construction have made it a bit of a dinosaur. The other problems associated with the trickling filter include the odors arising from contact with the wastes and psychoda flies. These little critters are nuisance organisms that live in the trickling filter and have a development life of about 2 weeks. They are very tiny and can be a great source of nuisance unless the filter is flooded for about 12–24 h on just less than a 2-week period. The technology that has replaced the trickling filter is the rotating biological contactor (RBC), which has its own limitations. The RBC is a series of slowly spinning Disks mounted on a shaft. The RBC does have its proponents who claim that it is more flexible than activated sludge, but one of its observed principal drawbacks is the fact that the Disks or rotors collect a biofilm (by design) and that adds enormous weight to the shaft. After a certain period these shafts develop stress cracks and snap, dropping the RBC into the wastewater tank.

Clarification for Biological Removals

Clarification will be handled in detail in the next chapter, but a few words of caution are important here.

The clarifier following an aerobic treatment process represents a separation of an active biomass from a liquid. There is a finite holding time generally not over 1–3 h in the clarifier. After that, the clarifier becomes anoxic and anaerobic decomposition begins where H_2S and N_2 gas are produced and where the clarifier is upset by gas bubbles.

The clarifier is easily overloaded. Conservatively designed clarifiers work best with low surface overflow rates (equivalent to average vertical bulk velocity expressed in flow units such as gallons per square foot per day (gpd/ft²/day). Clarifiers are generally built with an internal scraper arm mechanism and a surface cleaning mechanism, both of which very slowly rotate around a center shaft and sweep the settled solids, or the floating solids toward a collection point. Even this slow motion of the collector can cause horizontal currents, which upset the settling pattern in the clarifier.

Effluent weirs on clarifiers are also conservatively designed. There is some evidence that the loading rate of the effluent weir may be one of the most important features in developing good solids removals.

The purpose of a clarifier is threefold: (1) solids removal for recycle, (2) sludge thickening for wasting and recycle, and (3) removal of floating solids.

The clarifiers generally have an underflow or return cycle sludge concentration of less than 5% of the design flow (throughput or average daily flow through the plant), and this clarifier underflow is often more on the order of 1% to 2% of the daily design flow.

Other Solids Removals

In the early part of this chapter, we briefly addressed the issue of suspended material as a source of BOD or COD. In truth, the suspended solids loading to the biological treatment plant can comprise up to 50% of the total biological load applied to the treatment works. Depending upon the strength of the waste and the ability of the plant to handle the solids and maintain adequate aerobic treatment conditions, it may be necessary to have preclarification to remove the suspended material (and a portion of the substrate entering the plant).

When dealing with a domestic source, one can get everything from sand and clay particles to condoms, and footballs and bedsprings, and logs. In most processes, there is a provision for prescreening and solids size reduction to prevent the occurrence of the log, brick, or bedspring entering the treatment works. This is usually the function of a grit chamber. The grit chamber is a specific device designed to remove putrescible and non-putrescible solids in a fairly quick fashion. If there is no provision for grit removal, then the sand and other coarse solids will enter either the aeration basin or the first clarifier. Either way that will pose major problems for maintenance. The Sutro or proportional weir mentioned in the chapter on flow measurement is often used in a grit chamber, because it provides constant velocity through the chamber, regardless of the flow. This allows the heavier solids to settle to the chamber, where they can be removed.

Sludge Generation, Treatment, Storage, and Disposal

For all biological treatment plants and operations, the following general relationships hold (Fig. 6.5).

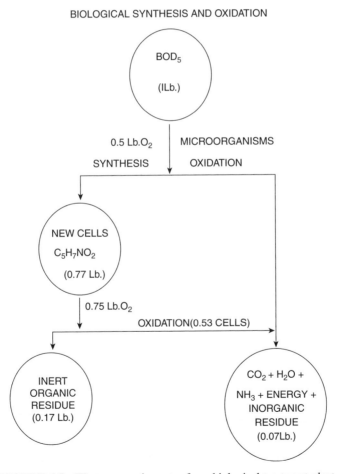

FIGURE 6.5   Waste generation rates from biological treatment plant.

One pound of BOD or organic matter yields about 0.77 lb of new cells and that requires about 0.5 lb of oxygen. As the food supply diminishes, the cells undergo a self-induced cannibalism. After some time, you will be left with 0.24 lb of inert organic residue.

Thus, if you have a waste where you are looking to remove about 100 lb of BOD per day, the approximate generation of biomass will be on the order of about 77 lb of solids per day. With time and digestion, that mass of solids will be reduced to 24 lb of solids. Unfortunately, those solids do not dewater well. If you are very fortunate, you will be able to collect them at about 18–24% solids on a dry weight basis, so that your 2.4 lb of solids will actually weigh about 100 lb give or take a bit.

The solids are collected from the underflow of a clarifier at between 1% and 3% solids. Depending upon the size of the treatment plant, the solids in the sludge may be thickened by stirring them for several hours in an anaerobic tank. The anaerobic stirring, called sludge thickening, will double the solids concentration. After that, the solids are conditioned further by the addition of all polymers and by centrifuging the sludge to concentrate it to between 10% and 13% solids.

The final solids concentration step is filtration. The sludge is processed through a belt filter press where the sludge is mechanically compressed and sheared in a traveling belt filter to attain a final solids concentration approaching 18% to 35%, depending upon the type of sludge and the processes used. The solids processing and disposal is one of the most costly operations in a wastewater treatment plant, especially when the sludge must be set to a sanitary landfill or, in rare cases, a hazardous-waste landfill.

BIOLOGICAL TREATMENT OF DIFFICULT WASTES

Not all chemicals are easy to treat. Biological waste treatment and oxygen demand were briefly discussed early in this chapter. A number of things cause difficulty with biological treatment of wastes. We have discussed shock loading and temperature effects, and biologically unbalanced loads. Now let us look at some other things that may cause difficulty in biological systems.

Each of these problems has a solution, but each is different in the solution.

Toxicity

Things that can cause toxicity include many of the following:

Metals: lead, antimony, copper, zinc, chromium, cadmium, nickel, manganese (permanganate), sliver,

Oxidizers: chlorine, chloramines, any of the group VII compounds in the periodic table, permanganates, ozone, fluorine, iodine, peroxides, and so on.

All of these compounds are direct toxins, because they directly interfere with the biological cycles in the cell and the cell enzymes.

Some organic materials are resistant because they are chlorinated, and chlorination makes them substantially harder to deal with. Others are toxic because they are phenolics. Phenol was the first major disinfectant. It can be biodegraded readily but it takes some work.

The point is that complex organic materials have some ability to biodegrade, if the conditions are correct. However, all biological treatment is as follows: The art of engineering a system so that the bacteria do what they will and want to do in a manner that coincides with your objectives. Stated in another way: "Given any combination of temperature, pressure, nutrients, and substrate, the bugs will do as they damn well please."[3] You have to understand what you are treating and how it degrades.

One of the best sources for information on biodegradability of all organic compounds is Karl Verschueren's Book, "*Handbook of Environmental Data on Organic Chemicals*," by Van Nostrand Rheinhold, NY. The book is quite complete and has excellent data on biodegradability for specific organic compounds. Much of the rate information in the book is unavailable elsewhere.

MODELING THE BIOLOGICAL PROCESS

In the Disk, which accompanies this book, there are several different biological models, which are usable and, which solve the basic differential equations for the growth of activated sludge. The first and oldest model is the SSSP model. It was developed in 1987 by Dr. Les Grady and one of his graduate students, Steven Birdrup at Clemson, SC. The program is an old DOS program but is highly flexible and runs a number of options and solves the basic equation of flow for nitrification and denitrification in wastewater systems, using the IAWQ model as a basis. It does run in DOS, and given the time of its development (1987), it is a very good work. The graphical interface is very rough by today's standards, but the price is right because it is free. It runs both static and dynamic simulations. In today's virtual DOS world, it is a little bit tough to run, but it has the advantage of being free. It is available from the following Web site: http://www.ces.clemson.edu/ees/sssp/.

---

[3]The sentence in quotations is humbly referred to as "Russell's law called waste treatment."

The program has been largely replaced by IAWQ's ASM1–ASM3 models, in the following sections as will be discussed. Both the SSSP and other models have been used by the author on a number of occasions, and all the models work quite well. The SSSP model is just a bit creaky by today's standards, and more recent work has changed a number of assumptions on how bacterial processes should be modeled.

STEADY

The second free model is the STEADY model. It was developed at the University of Texas, and it is a self-installing zip file of about 3.7 MB. The author, Dr. Gerry Speitel, posted the model on the Web at: http://www.ce.utexas.edu/prof/speitel/steady/steady.htm.

The model provides a static solution to a plant design and allows one to set up their own simulation and run it to develop a steady state design solution. Unfortunately it is not a dynamic model. In this writing, there are no ASM2, ASM2d, or ASM3 models.

The model will allow one to configure a wastewater treatment plant for certain limited designs and develop some data on the plant. It also has some good graphics and a well-defined adjustable interface and good screens for a manual. The principal limitation is that it does not model the clarification or the refinements of the activated sludge model very well. It is fun to play with and considering the price, it is well worth learning about. Because it is simple, it can be used in English or metric units.

A sample of the steady screen shot is shown in Figure 6.6.

JASS

A third free model is the JASS model that was developed by Uppsala University (Sweden). The model is in Java and is available on the Web site only. The Web address is http://user.it.uu.se/~psa/. A graphic of the treatment plant graphics is shown in Figure 6.7.

The model appears to have some flexibility, but it is clearly a students' tool and has a list of bug fixes and other associated changes. The system will provide some nitrogen control but lacks flexibility in the design process for configuration. Even the laboratory model has the same limitations. One of the principal drawbacks of using it is the possibility that someone, namely a student programmer or a professor, may have modified it and not documented his changes. It provides reasonable results but with a fixed configuration.

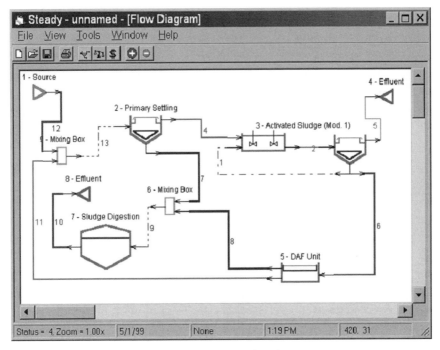

FIGURE 6.6   Screen shot of study program.

For the professional, there are some very good models around, but they come with a price. Most, if not all, of the models are from Europe or Canada.

SCILAB/SeTS

The SeTS (Sewage Treatment Simulation) runs under Scilab, a free commercial program from the University of Karlsruhe. It is a GNU-licensed wastewater simulator. It has models for ASM1–ASM3 and ADM 1 (Anaerobic Digestion Model #1). The Web site is: http://www.uni-karlsruhe.de/~gh31/SeTS.

The interface appears reasonable though somewhat clumsy.

AVAILABLE COMMERCIAL MODELING TOOLS

Process Advisor

Process Advisor is quick but somewhat cumbersome. The demonstration version is adequate, but it is focused more on operations than on design. It is not a predictive tool for design but more on the order

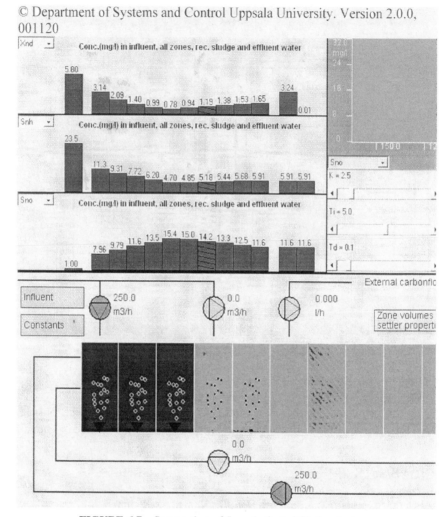

FIGURE 6.7 Screen shot of Java Activated Sludge System.

of a data mining of past operations and "fixes" for previous problems solved.

Hydromantis and GPSX

GPSX is produced by Hydromantis, Inc. in Hamilton, Ontario, and has a number of models for activated sludge and treatment works modeling, including: ASM1, ASM2, ASM2d, ASM3, and a temperature-dependent version of the ASM1, and an Anaerobic Digester Model ADM1. GPSX links directly to and from their Capdet Works program for cost estimation and can

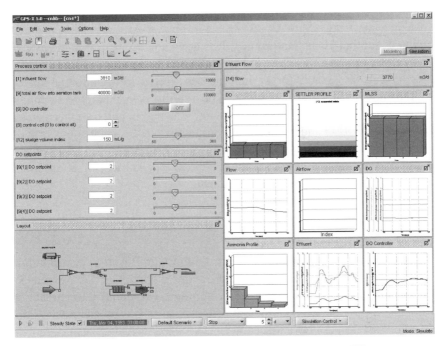

FIGURE 6.8 Screen shot of hydromantis software GPSX.

perform dynamic simulations for volatile organic compounds, and metals by linking with their TOXCHEM program. A sample of the screen model is shown in Figure 6.8. Their Web site is: http://www.hydromantis.com. The program is expensive of the order of about $17,500 a full license. The base model sells for approximately $4900, and an advanced single user license is $11,500. GPSX comes with 60 preconfigured modules to make the setup of the wastewater models easier (Fig. 6.8).

Matlab

Matlab that has an annual license fee of US $2500 per year for a single user also has an activated sludge simulator. It is widely used by the universities to develop their research, and many programs have been written in it. The best advice is to ask questions and thoroughly investigate the system. Sometimes this information does not appear on the information provided in a company's list of operating requirements for their software. Matlab is an extremely powerful mathematical system, and it can solve second order differential equations, which are well beyond the scope of many of the other commercial wastewater modeling programs.

Biowin

Biowin is another Canadian product. It is good and seems to run fairly well. It was designed by engineers and is relatively straightforward in use. The Biowin Web site has a lot of helpful descriptive information about the program and the technical features. The pricing is comparable to that of GPSX. A full release is about $20,000, and it is sold in parts, so that you can start at about $7000 and go upward from there.

Biowin is extremely popular and in wide use in a number of consulting shops. They have most, if not all, the IWA developed models in addition to a number of proprietary models of their own.

STOAT

STOAT is sold under commercial license from Water Resources Corporation, Ltd (WRc, Ltd) in England. Both the programs were designed by engineers, and the author has used STOAT for designing commercial water treatment programs. STOAT is a very efficient program, as it was created in Fortran.

The output from STOAT was to an Excel file exclusively. The program did all the simulation work at one time and produced a dynamic output file with all the information asked for. The programs were relatively easy to use, and the help manuals are exhaustively complete as is the technical information. Their latest training manual is over 400 pages and very thorough.

WEST

The World-wide Engine for Simulation Training and Automation (WEST) software is produced by Hemmis. The interface is very good and impressive. The WEST product produces a dynamic model output that has an outstanding graphical interface and true dynamic environment. The program is by far the most flexible and adaptable program in the market, and it has the option of being programmable so that new variables can be defined during the setup, the models modified, and new models incorporated. The progress of these new elements can be tracked during execution. Most of the models, with some few exceptions, are open source and open code so that they can be modified to tailor their work to specific applications.

The program also has a feature of tunable parameters or "sliders" that allow dynamic control of the model parameters during execution. WEST was put together with the assistance of the Biomath department of the University of Ghent. As a matter of personal opinion, I prefer WEST because I am more familiar with it. The price structure for WEST is comparable with that of GPSX and Biowin.

WEST has the following models available:

- **ASM1** Includes carbon and nitrogen removal.
- **ASM2** Includes carbon, nitrogen, and phosphor removal.
- **ASM2d** Includes carbon, nitrogen, and phosphor removal. This is a modification of ASM2 in order to improve the accuracy of the predictions.
- **ASM3** Includes carbon and nitrogen removal. This is a modification of ASM1 in order to improve the accuracy of the predictions.
- **ADM1** Anaerobic digester model
- **Buffertank** (variable volume with a weir, variable volume with a pump, fixed volume)
- **Primary clarifier** (pointsettler, Otterpohl and Freund, Takacs reactive...)
- **Activated sludge unit** (plug flow, oxidation ditch, fixed/variable volume)
- **Sequencing batch reactor** (SBR pointsettler, SBR multilayer)
- **Secondary clarifier** (pointsettler, Otterpohl and Freund, Takacs, Marsili Libelli)
- **Anaerobic digester** (Siegrist model for an anaerobic digester), and IWA's Anaerobic Digester Model ADM1
- **Trickling filter** (Rauch)
- **Generator** (block, sinus, double sinus—used to generate influent files)
- **River quality** (bulk benthic, river model No. 1)
- **Sensor** (flow, DO, NO_3, NH_4, PO_4, TSS, COD, BOD, TP)
- **Controller** (P, PI, PID, OnOff, Ratio, Saturation)
- And several other models, in addition to analysis software that allows one to do parameter estimation, curve fitting, scenario analysis, and error analysis with confidence intervals on the data.

A sample of the WEST model building screen and sample output screens are shown in Figure 6.9.

These simple graphical tools allow you to quickly adjust the input variables using your computer's mouse. The advantage is that it saves time and tedious input when one is experimenting with various designs. Their data input and output can be from a comma-delimited or tab-delimited file, and the output can be graphical or numerical depending upon your preferences.

---

*Author's Note: The author has a business (sales) relationship with Hemmis, NV the publishers of WEST, and as a result, the frequent references to WEST represent both his personal preferences and familiarity with their product. However, he has attempted not to make the following material, because of its breadth, a commercial for WEST, but an attempt to discuss modeling using WEST as an example.

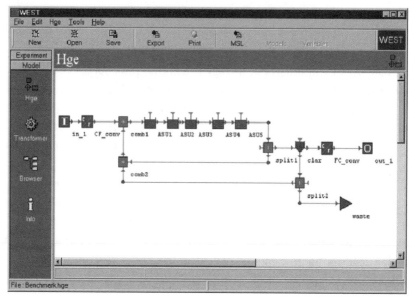

FIGURE 6.9 WEST software typical plant configuration.

Hemmis is producing a free download executable model. The model can be found on their Web site: http://www.hemmis.be

Two samples from WEST are shown in Figures 6.10–6.12. The first is an SBR configuration, the second is a complex multiple tank and multiple control system, and the third is a sample output from a user interface from a recent paper.

One of the strongest features in WEST is the ability to implement and automate process control strategies for plant automation. Several European wastewater treatment plants are using WEST as a control and data aquisition system as well as predictive model.

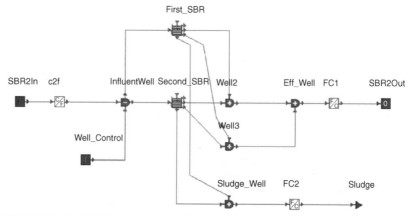

FIGURE 6.10 WEST configuration for a two tank sequencing batch reactor system.

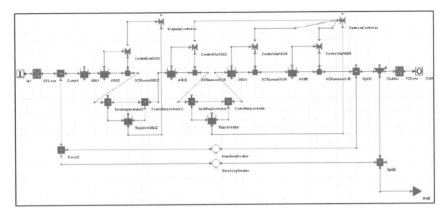

FIGURE 6.11   WEST configuration for multitank system with respirometry control.

MODELING GUIDANCE

If you are contemplating plant design or plant operations, I believe that the future is in using simulations to provide your plant with models for operations. Only in that way ultimately you will be able to get the wastewater treatment plant design and operations from the dark ages into the twenty-first century. If you are going to design a plant or modify a plant, plan on using someone's software for designing or modeling.

Modeling and simulations require a complete new way of thinking about process considerations. The data input for even a simple simulation using ASM1 can be considerable. For example, consider the following taken from

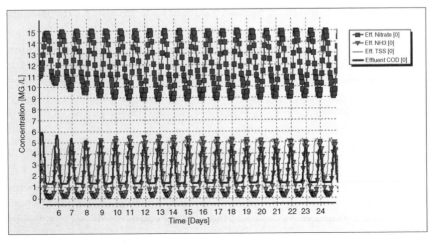

FIGURE 6.12   WEST example of WEST dynamic control output graphics.

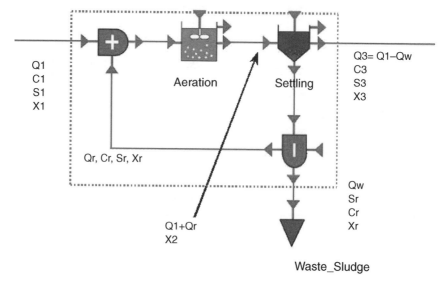

Q1
C1
S1
X1

Aeration

Settling

Q3= Q1–Qw
C3
S3
X3

Qr, Cr, Sr, Xr

Q1+Qr
X2

Qw
Sr
Cr
Xr

Waste_Sludge

FIGURE 6.13   WEST graphics: Basic schematic of activated sludge system.

the WEST models for ASM1. The WEST nomenclature is extremely close to the original ASM1 development nomenclature (Fig. 6.13).

First Thoughts

Start back with the description of the modeling and parametric elements on an activated sludge plant. Let us begin again with a slightly different nomenclature.

The models for activated sludge and biological processes utilize COD. We can use BOD, but it has to be transformed into COD and the COD data need to be apportioned into respective components.

Wastewater contains all sorts of substances. The principal one is carbon. The carbon is transformed to CO_2 by bacterial action. Other compounds are converted to cell growth, nitrate, sulfate, and so forth. However, these depend upon the conditions and degree of biological activity. Aerobic conditions lead to oxidized metabolites. Anaerobic conditions lead to ammonia, H_2S, and various types of organic acids-reduced organics.

Because wastewater contains a variety of compounds, we will start by focusing on the carbon variable. That is the first principal and the one used by all the models. It must also be noted that the commercial models such as WEST and all the others use a number of equations to track the components of wastewater. When one considers the various components and bacteria in wastewater, it is possible to get at least 13 first-order differential equations and when you add things like recycle flows and clarifiers, it is possible that for a simple configuration, the number of differential equations can easily reach up to 40 or more. The only practical way to solve these equations is by using a modeling system.

How to Use the Matrix and the Equations

Given the plant depicted above and an effluent concentration of 10 mg/l COD, we will consider a design flow rate of 5 million gallons per day, which is a Q of 3.785 L/Gal $\times$ 5 MGD $= 18.925$ M L/D $= 18925$ M^3/D at a strength of 500 mg/l $(= 0.5$ kg COD (S)/M^3. The plant has no recycle. Find the aeration tank volume.

We have the following information provided:
$r_{xs} = 3$ kg COD (S)/kg(B) COD 3 kg of substrate consumed per kilogram of bacteria, and yield for substrate is 0.4 g COD(B)/kg COD (S) $- 0.4$ kg COD of bacteria per kilogram of substrate consumed.

Look at the equations above and find the necessary volume for the tank

$$V = [Q_1 C_1 - Q_3 C_3]/[r_x X_b]$$

and

$$X_b = Y(C_1 - C_3)$$
$$X = 0.4(500 - 10) = 0.4 \times 490 = 196.0$$

and

$$V = [Q_1 C_1 - Q_3 C_3]/[r_x X_b]$$
$$= (18925 \times 500 - 18925 \times 10)/(3 \times 196) = 15770.83 \text{ m}^3$$

Now if the plant has recycle, $Q_w > 0$.

We can go on and look at the overall process doing mass balances in any of the number of ways and at various points. The critical element is to take the internal tank reactions and the recycle rate into consideration.

Performing a mass balance around the aeration tank, and using that to calculate the tank effluent concentration, we get the following:

$$X_1 C_1 + X_r C_r + (Q_1 + Q_r)(C_{ATIn} - C_{ATOut})Y = (Q_1 + Q_4)X_2$$

Obviously that requires the knowledge of a bit more information. The overall substrate removal rate for aerobic (heterotrophic) growth is as follows:

$$r = (\mu/Y)(S_2/(S_2 + K_s))(S_{O_2} + K_{SO_2})X_2$$

where S_2 is the concentration of organic matter in the aeration tank. The mass balance for the entire plant is

Input	+	Hydrolyzed	−	Removed	=	Output
$Q_1 S_1$	+	$r_x V_2 v_{xs}$	−	$r_{vs} V_2$	=	$Q_3 S_3$

where v_{xs} is the hydrolysis coefficient.

All the above is for a simple set of reactions, but it forms the basics for the modeling.

THE IWA MODELS FOR ACTIVATED SLUDGE

The IWA models currently in use for modeling the activated sludge process are ASM1, ASM2, ASM2d, and ASM3. ASM1 is the oldest, dating from about 1987, and ASM3 is the newest, which generally replaces ASM1. ASM3 is not as widely used, but because it is easier to use, it may become much more popular in the near future. The problem with ASM1 is that it is difficult to fractionate the influent waste stream in the manner required for the model without a lot of complications and possibly at the expense of sampling and analysis. The ASM1 and ASM3 models can be used to model nitrate removal.

ASM2 and ASM2d are written with the phosphorous variable in mind. They are re-written rather to consider the fact that activated sludge comprises cellular biomass that has the ability to store and use phosphorous. ASM2d is specifically for phosphorous removal, and it also considers the stoichiometric addition of metal salts from an exterior source as a method of removing phosphorous.

The ASM models are written in the same matrix notation used in Table 6.8, and in the simplest model, ASM1, there are eight rate equations: Aerobic growth of hetrotrophs, anoxic growth of hetrotrophs, aerobic growth of autotrophs, anoxic growth of autotrophs, decay of hetrotrophs, decay of autotrophs, ammonification of soluble organic nitrogen, hydrolysis of entrapped organics, and hydrolysis of entrapped organic nitrogen. So eight equations, with 13 variables all expressed in matrix form.

The use of the words, heterotrophs and autotrophs, refers to the wastewater bacteria, which are capable of using exterior or interior sources of carbon to oxidize the wastewater.

Heterotrophs are assumed to be the utility organisms, capable of doing a wide variety of things in a biomass system. They grow aerobically, anoxically and may be active in anaerobic fermentation. They are responsible for hydrolysis of particulate substrates and can use all degradable organic substrates under all conditions.

Autotrophs are nitrifying organisms, which are responsible for nitrification. They are obligate aerobes, classified as chemo-litho-autotrophs and are responsible for ammonia oxidation to nitrate (nitrosomonas and nitrobacter).

Name	Description
H_2O	Water
S_I	Inert soluble matter
S_S	Readily biodegradable matter
S_O	Dissolved oxygen
S_NO	Nitrate and nitrite
S_NH	Free and ionized ammonia
S_ND	Soluble biodegradable organic nitrogen
S_ALK	Alkalinity
X_I	Inert particulate matter
X_S	Slowly biodegradable matter
X_BH	Heterotrophic biomass
X_BA	Autotrophic biomass
X_P	Particulate products resulting from biomass decay
X_ND	Particulate biodegradable organic nitrogen

These components are also used to characterize the influent of the wastewater treatment plant.

Additionally, the following parameters are also required for the specification of the state variables in the modeling process (Table 6.10).

Y_H and **Y_A** must be larger then zero. (**Y_H** > 0 and **Y_A** > 0).

In some of the other supplemental materials on the Disk there is information about the preparation of the ASM1 model and the preparation of the influent file. Modeling is not necessarily easy or fast, and it can take a few days to set up and run a specific complex configuration, but it is worth the trouble to do it correctly.

TABLE 6.10   Sample of Modeling Constants Formulation in WEST

Variable	Description	Units
Y_H	Yield for heterotrophic biomass	g COD/g COD
i_XB	Mass of nitrogen per mass of COD in biomass	g N/g COD
Y_A	Yield for autotrophic biomass	g COD/g N
f_P	Fraction of biomass converted to inert matter	—
i_XP	Mass of nitrogen per mass of COD in products formed	g N/g COD
K_S	Half-saturation coefficient for heterotrophic biomass	gCOD/m^3
K_OH	Oxygen half-saturation coefficient for heterotrophic biomass	gO$_2$/m^3
K_NO	Nitrate half-saturation coefficient for dentrifying heterotrophic biomass	gNO$_3$-N/m^3
b_H	Decay coefficient for heterotrophic biomass	1/d
mu_H	Maximum specific growth rate for heterotrophic biomass	1/d
n_g	Correction factor for anoxic growth of heterotrophs	—
-K_OA	Oxygen half-saturation coefficient for autotrophic biomass	gO$_2$/m^3
K_NH	Ammonia half-saturation coefficient for autotrophic biomass	gNH$_3$-N/m^3
b_A	Decay coefficient for autotrophic biomass	1/d
mu_A	Maximum specific growth rate for autotrophic biomass	1/d
k_a	Maximum specific ammonification rate	m^3/(gCOD.d)
K_X	Half-saturation coefficient for hydrolysis of slowly biodegradable substrate	gCOD/gCOD
k_h	Maximum specific hydrolysis rate	gCOD/(gCOD.d)
n_h	Correction factor for anoxic hydrolysis	—
Kla	Oxygen transfer coefficient	1/d
S_O_Sat	Oxygen saturation concentration	g/m^3

With regard to the model parameters, the accepted values are published in the IWA model documents and will be constant for many waste streams; the differences in some of the reaction constants may be slight and insignificant. For really accurate simulations, an effort must be made to perform sampling, collect the influent data, and then assign the correct fractions to the influent.

The actual description of the procedure for modeling a plant is also a subject one can spend hours and days discussing and learning.

7

PRECIPITATION AND SEDIMENTATION

Theory of sedimentation
Clarifiers and their design
Lamellas and specialty devices

THEORY OF SEDIMENTATION

The basic theory of sedimentation is a balance between gravity and drag forces.

Particle forces, $F_1 = (\rho_s - \rho)gV$ Impulse or gravity forces

Drag forces, $F_d = C_D A \rho V^2 / 2g$

R is the Reynolds number

For spheres up to $R = 10{,}000$ $\quad C_D = 24 + \dfrac{3}{R} + \dfrac{0.34}{\sqrt{R}}$

At steady state, $F_1 = F_d$ and this reduces to Stoke's law for values of R less than 1

$$V = \frac{g}{18} \frac{(\rho_s - \rho)}{\rho} \frac{d^2}{\mu}$$

This is the basic formula for determining the settling rate for most discrete solids. Of course as the size and density increase, the drag forces also

Practical Wastewater Treatment, by David L. Russell
Copyright © 2006 John Wiley & Sons, Inc.

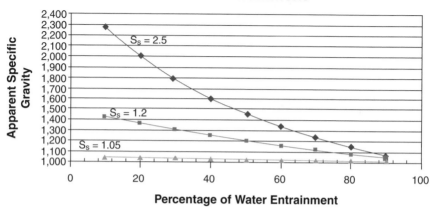

FIGURE 7.1   Change in specific gravity of a particle with entrapped water.

increase. The keyword here is discrete. If the solids flux is too high, and/or the solids have a tendency to be "sticky" or agglomerate, settling can be hindered. Concentrations of "sewage" or organic biosolids above 3500 mg/l and silt above 6000 mg/l can hinder settling. Often the result is a decrease in either the bulk settling rate or zone settling rate.

One of the reasons for hindered settling is floc agglomeration and water entrapment, which leads to reduced apparent density (see Fig. 7.1).

$$S_m = \frac{100}{(100 - P)/S_s + P/S}$$

where S_m = apparent specific gravity of a group of particles; p = percentage of water entrained; S_s = true specific gravity, and S = specific gravity of liquid.

CLARIFIERS AND THEIR DESIGN

Clarifiers are available in two shapes, round and rectangular. Depending upon specific densities, most biological flocs move at rates of 1–2 m/h. Each clarifier has an inlet zone, an exit zone, a dragout or collection device, and a sludge withdrawal area. Clarifiers are moderately well understood and can be modeled, but overall there is a lot of lore and practice that is embodied in design codes and that has been established by trial and error. One year everyone will rush to put in a specific type of collector arm, the next year it will be a modification to the inlet structure or the outlet structure or both. A typical clarifier is shown in Figure 7.2.

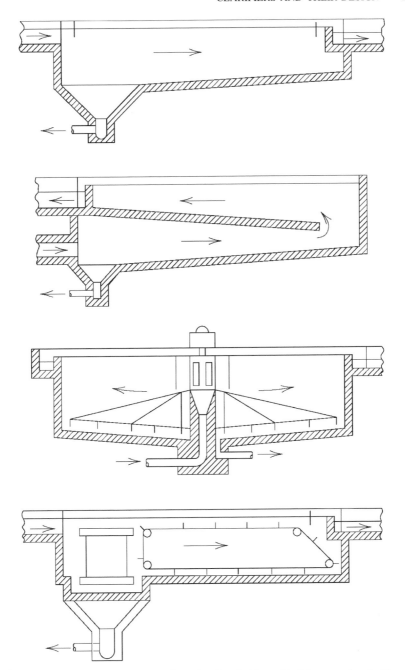

FIGURE 7.2   Typical design and configuration for clarifiers. (Top) Rectangular clarifier with gravity drainage for sludge. (Top middle) Compact rectangular clarifier which is deeper but has longer settling path. (Bottom middle) Center feed circular clarifier with submerged sludge scraping mechanism. (Bottom) Rectangular clarifier with a chain dragout and sludge scraping mechanism. Figure 7.2 is from Fair and Geyer, *Water Supply and Wastewater Disposal*, McGraw Hill, 1964.

There are a number of good practices that have been hinted at in a clarifier design, especially when one looks at biological clarifiers.

Bulk Velocity – Surface Loading Rate

Make sure that if you are using an upflow clarifier, the bulk velocity of the clarifier is not greater than the settling rate of the smallest particles you want to get out. This is often measured by overflow rate, and has been embodied in many codes, including the TEN STATES STANDARDS.[1] Bulk loading rates are defined in terms of gallons per square foot per day. This is a velocity term – equal to about 33 $m^3/day/m^2$ – at a maximum of 800 gallons/ft^2/day. This is equivalent to about 1.5 m/h as a floc settling rate. For all intents and purposes for a biological floc, it is a good maximum number and should not be exceeded without good basis for the use of other criteria.

Hydraulic Detention Time

This is often set arbitrarily at 2–3 h, without good reason. It depends upon the biological activity of the sludge and the MLSS in the clarifier feed. When one is dealing with a biologically active sludge and low to moderate dissolved oxygen entering the clarifier, it is often better to keep the detention times shorter. Again, solids con- centrations, the oxygen in the water, and the oxygen uptake rate should govern the selection of this parameter. It is never a good idea to let the clarifier become anoxic or anaerobic.

Solids loading rate should not exceed 20 lb/day/ft^2 of surface area (98 kg/m^2/day). This is not necessarily true. Depending upon the parameters of the sludge, a number of clarifiers have operated 50% or more over the rate. The quantity of solids is more a function of how easily the settled solids can be moved for collection and removal from the clarifier.

The concentration of the sludge is controlled by the settling characteristics and the withdrawal rate.

Many settling tests and measures of sludge settleability are conducted improperly in design and in evaluation. The principal problem is that they are conducted in a graduate cylinder of relatively small diameter where edge effects occur and give false readings.

---

[1]Upper Great Lakes Regional Board of Sanitary Engineers developed these standards, and they have seen almost universal application.

FIGURE 7.3 Circular clarifier under construction—Observed the center baffle that is used to direct the flow downward, under the baffle, before the flow flows outward to the launder ring (scum and floatable barrier) and the overflow weir around the periphery. *Source:* Greeley and Hansen-Lafayette, Ind.

Weir overflow rates should not exceed 10,000 gallons/day/ft of weir or 11.54 m³/day/m. Lower rates are preferable because the local velocities can draw local solids out of the system.

Do not forget that the clarifier throughput rates should include the recycle rate from the primary aeration system. Design accordingly.

Figure 7.3 shows how some of the innards of a clarifier look like.

The innards and the sludge scraping arrangement from a sludge thickener are shown in Figure 7.4.

The thickener concentrates the clarifier under flow by a factor between 3 and 5.

LAMELLAS AND SPECIALTY DEVICES

Lamellas

Lamellas are a special case for a tube clarifier. The theory of design is the same, but the internal plates give the sludge a shorter distance to travel, and

FIGURE 7.4   Innards of a sludge thickener. Note steeply sloping sides and mechanical rake which promotes sludge compaction & collection in the center well. Other manufacturers will have tanks with steeper sides and vertical poles (rakes) on the collector mechanism to promote thickening. Photo courtesy Wes-Tech Engineering Salt Lake City, UT.

they are generally more efficient. However, it may be very difficult to clean them if there are problems with the sludge.

A lamella is shown in Figure 7.5.

Density currents can cause efficiency reductions in a clarifier. They are caused by (1) eddy currents, (2) wind induced currents in the settling, and (3) convection and density currents. This is called damping.

The effect of damping is shown by the effect of increasing the number of plates and decreasing the distance the particulate materials have to settle

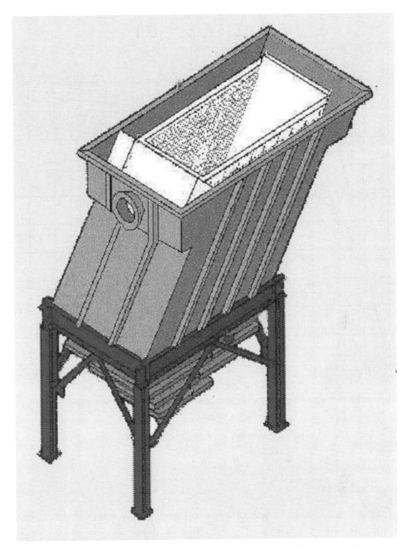

FIGURE 7.5   Drawing of a Lamella (by Parkson Corporation).

before they encounter a plate (Figs. 7.7 and 7.8). Damping follows the general formula given below.

If Y_o is the orignial settling velocity, and Y is the apparent settling velocity, then $Y - Y_o$ is the amount suspended matter of settling velocity V_o still in solution at time t.

For a specific clarifier, there is the following overall relationship:

$$(Y/Y_o) = 1 - [1 + V_o/(nQA)]^{-n}$$

$$\frac{y}{y_0} = 1 - \left(1 + \frac{1}{n}\frac{v_0}{Q/A}\right)^{-n}$$

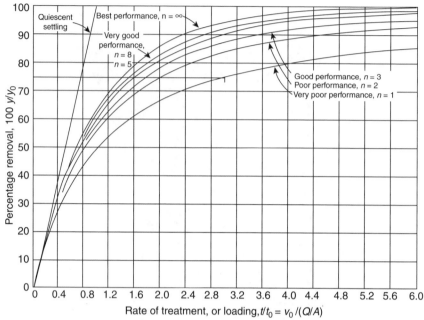

Performance curves for settling basins of varying effectiveness.
After Hazen.

FIGURE 7.6   The increases in clarifier performance owing to damping of eddy currents. *Source:* From Fair and Geyer: op. cit.

Membrane Filters

One of the most exciting developments in the past few years is the utilization of membrane filters in lieu of a clarifier. This is a unique development because despite their higher initial cost, either in fixed or in flexible form, they have a superior performance over clarifiers at modest head losses. The head losses are between 3 and 5 psi across the membrane, but the real advantage is that they have a long operation life of 10 years or more, and an effluent TSS, which is zero.

Translated into operations terms this means that the effluent TSS is always less than 1 mg/l because the membrane only passes solids with a size of less than 0.45 μm, too fine for most tests to pick up. The other advantage is that the solids wasted are the only ones lost, at the discretion of the operator.

If the membranes get plugged, the backwash is with acetic acid or other mild acid. A picture of the membrane effluent filter is shown in Figure 7.8.

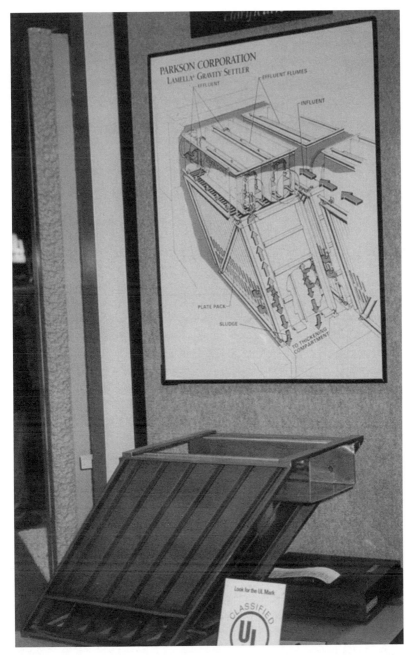

FIGURE 7.7   Lamella model and drawing by Parkson taken at WEFTEC'03. A Lamella by the Parkson Corporation

FIGURE 7.8 Spaghetti strand hollow tube membrane filter clarifier.

8

FILTRATION THEORY AND PRACTICE

Depth filters design: Theory and practice
Filtration hydraulics
Hydraulics of filter washing
Skin filters
Filter elements and design

DEPTH FILTERS DESIGN: THEORY AND PRACTICE

There are several types of filters in the marketplace. The most popular appear to be the "inverted" or "mixed media" filter, which employs various filter media of different densities and sizes to get the filtration, and the sand filter, which uses sand and gravel of different sizes to construct the filter. The filters are built in reverse from each other as shown in Figure 8.1.

The mixed media filters utilize varying densities and sizes of media to achieve the mixed media effect. In the conventional sand filters, the media density is about 2.65, while in the mixed media filters, the densities range between 1.5 for plastic and artificial media and 4–4.5 for garnet sands and corundum sands. This gives the media a reverse gradient and allows deeper penetration of the solids in the filter (see Fig. 8.2).

Sizing of Filters by Flow Rate

There are three general classifications for sand filters: rapid sand filters, slow sand filters, and pressure sand filters. All three are built along the same general configuration for a conventional media filter shown below (left

Practical Wastewater Treatment, by David L. Russell
Copyright © 2006 John Wiley & Sons, Inc.

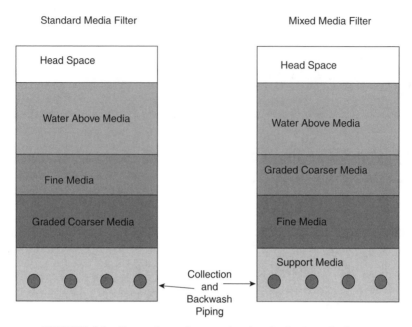

FIGURE 8.1   Comparison of conventional and mixed media filters.

drawing). The principal difference between the three is the flow rate and the pressure drop across the filter. The slow sand filter is sometimes used in municipal water supplies and has a flow rate of under 2 gallons/minute/ square foot (face velocity of 4.89 m/hr) . The conventional sand filter has a flow rate from 2-6 gallons per minute per square foot, (4.89-29.5 m/hr) and

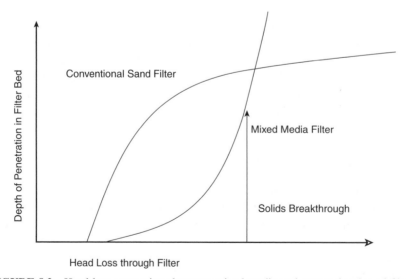

FIGURE 8.2   Head loss comparison between mixed media and conventional sand filters.

TABLE 8.1 Comparison of Filter Types

Filter Type	Filter Rate Conventional (gallon/ft²/day)	Metric L/M/Sq. M.	Use	Bed Depth (cm)	Grain Sizes (mm)	Uniformity Coefficient	Density of Media	Head Losses (M)
Slow Sand Filters	20-100-250	0.6-7.1	Final polishing filter after bio treatment	30 gravel 100 sand	0.25-0.35	2-3	2.65	0.2 initial 1.5 final
Rapid Sand Filters	3000-7500	80-250	Conventional solids removal in all types of applications	50 gravel 80-100 sand	0.45+	1.5+	2.65	0.3 initial 3-5 final
Mixed Media Filters	7500-22000	250-700	High rate filter applications	50 garnet sand 30 coarse sand 30 plastic media	0.45+	1.2+	4.5 2.65 1.8-2.2	Varies with Flow rate
Skin Filters	3000-7500	80-250	Swimming pools, commercial filtration Requires precoat media Has low backwash requirement	0.3 cm max used cloth or wire as support	0.05+	N/A	3-Feb	Varies

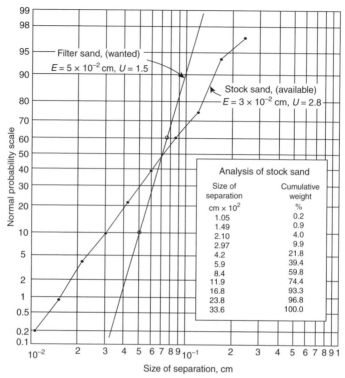

FIGURE 8.3 Sizing of typical filter sands.

the pressure sand filter has a flow rate of greater than 8 gallons per minute per square foot. The differing flow rates and pressure drops all impact the solids removal and the physical configuration of the filter, including the type of vessel and the backwash appurtenances used in the filter. Table 8.1 presents useful information on various types of filters in a slightly different format.

Uniformity Coefficient and Effective Grain Size

Effective grain size is the size of 10% of the smallest media, or D_{10}. Uniformity coefficient (U_1) is the ratio of D_{60}/D_{10}.

FILTRATION HYDRAULICS

For general hydraulic losses through a filter, the following equation holds:

$$h/L = 1.067(C_d v^2)/(g f^4 d)$$

where C_d = Drag coefficient; g = acceleration of gravity; v = face velocity of liquid; and f = porosity of the filter bed (expressed as a decimal).

At laminar conditions, the equation becomes

$$h/L = 25.6(vv)/(gf^4d^2)$$

where v is the kinematic viscosity of the working fluid.

HYDRAULICS OF FILTER WASHING

For granular media filters, the filter bed is set by backwashing. The bulk upflow velocity through the bed and the settling rate of the particle determine which particles will be raised and by how much. Typically sand beds expand between 75% and 100% during backwash, and less with mixed media filters. However, both types must be scoured or violently agitated during backwash to break up mud balls and accumulated agglomerated masses.

One of the most useful things encountered in backwash hydraulics for a filter is that the technique used to insure a uniform distribution of the backwash water on the underdrain system for the filter bed is also useful for many environmental problems where it is important to have uniform flow distribution over a long distance.

The problem is a curious one, but the solution is relatively simple.

Question: How do we insure that in a pipe with holes in it, we get uniform distribution of the flow either into or from the pipe along the entire length of the pipe?

Answer: By uniform sizing of the distribution losses so that the nozzle or opening is significantly greater in head loss than the remainder of the head loss through the pipe gallery. This is useful in controlling the discharge into a pipe as well as the discharge across a filter gallery, and it can be used for horizontal wells and other similar problems where even distribution is required through a long pipe. This solution has even worked in distribution and collection systems for long pipes (~130 m) where even flow was required.

For granular media filters, the backwash is between 2% and 10% of the throughput. For skin filters, it is generally well under 1%.

Filters have their own limitations. The pore opening dictates the efficiency with which solids can be removed from a liquid. Figure 8.4, illustrates the various ranges of solids, and the types of separation techniques, which must be applied to remove them.

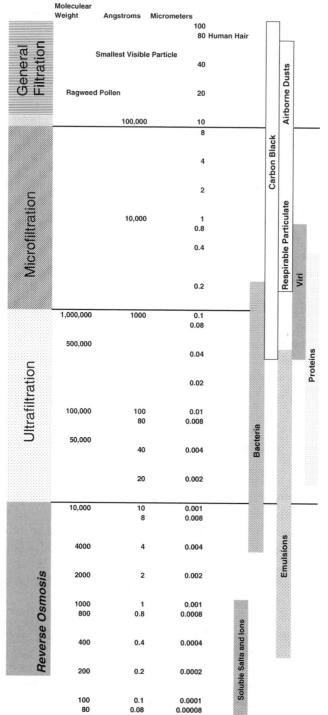

FIGURE 8.4 Some data on particle sizes relative to removals by filtration systems.

SKIN FILTERS

Skin filters are called so because they are thin, and most of the action in a skin filter takes place by physical plugging of the filter openings. When the openings are plugged, it is time to change or backwash the filter. Think of a filter press in this regard.

Skin filters are everywhere. In the chemical industry, one of the most common uses is the filter press. In the automotive industry, two of the most common types of filters are the fuel filter, which is located in the carburetion system and the oil filtration system. In the residential market, one of the most common uses of a skin filter is for filtration of the water in a swimming pool.

There are two types of skin filters in common use: Precoat and non-precoat or disposable filters. The precoat filter uses a specific media, which is the support structure and the filter as well. This medium, usually a filter cloth, is most often a polypropylene, or polyethylene fiber—but many other fibers are also in wide use. The fibers are usually very close-weave, with pore openings under $100\,\mu$ (sieve sizes of number 140 and higher). Depending upon the strength of the fiber, and the face velocity (pressure loss across the medium), the filter cloth may be supported by a underlayment structure. When the filter is in place, and the filter chamber flooded, a recirculating body feed is added to develop a precoat. This precoat often consists of diatomaceous earth. The purpose of a precoat is to decrease the effective pore size of the openings in the filter cloth and increase the efficiency of the filter (Fig. 8.5).

During the course of the filter cycle, the precoat needs to be enhanced with a body feed to maintain the porosity of the body feed. This body feed is generally a slurry of 1–2% solid, which is metered into the filter feed. The rate is dependent upon the nature of the solids being filtered, the pressure drop, and the filter rate. Body feed can enhance filter cycles from 10% to 50% over nonbody feed filtration.

Most often the filter aide is diatomaceous earth (filter material, or an expanded granular material such as lava, which is crushed and then heated

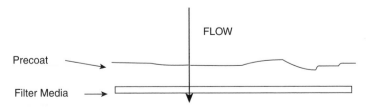

FIGURE 8.5 Precoat layer on a skin filter.

to expand). Most times, the filter body precoat is developed on the basis of testing by the manufacturer, but generally it amounts to about 0.25 cm. One of the most common materials is *diatomaceous earth*. The following description taken from *Material Safety Data Sheets* highlights the properties of diatomaceous earth.

Diatomaceous Earth, DE

Synonyms: *Diatomaceous silica, diatomite, de, & kieselguhr*
Designations:

Chemical Name: *Diatomaceous silica*
Chemical Formulas: $SiO_2 \cdot nH_2O$

General Description:
A naturally occuring mineral derived from microscopic in size fossilized remains of marine diatoms. It has high absorption, low bulk density and high brightness.

Typical Chemical Purities Available:

There are two basic grades available.

One grade is suitable for use as a garden insecticide (crystalline silica content is around 0.36% to 1.12%) and is usually approved by both the EPA and the FDA.

The other grade is sold by swimming pool suppliers (crystalline silica content is close to 60%) as a filtering agent. $SiO_2 = 86.30\%$, $Al_2O_3 = 4.50\%$, $Fe_2O_3 = 1.57\%$, and $CaO = 1.43\%$.

Typical Granulations Available:

Powder sized finer than 45 microns

Nominal Physical Constants:

Dry density (lbs./ft^3)	*9.5–13.0*
Apparent wet density (lbs./ft^3)	*20.0–27.5*
Specific gravity	*~2.0*
G. E. brightness	*64–92*
Melting point (°C)	
Boiling point (°C)	
Surface area (m^2/g)	*30*
L.O.I. (%)	*4.00*
Moisture (%)	*~ 1.0*
pH (10% slurry)	*7.0–10.0*
Fusion point (°C)	*1715*
Color	*Off-whitte to pink*
Refractive index	*1.46*

Other Names: *diatomaceous earth diatomaceous silica diatomite kieselguhr.*

Typical Applications:

A silica source in the production of calcium silicates, insulation bricks and material in safes, fireproof filing cabinets, etc. Used in the paint, varnish, lacquer, and polish industries. Used as an insecticide in gardens and in swimming pools as a filtering agent.

Packaging Options: *Bags, drums and bulk bags*

The application rate for filtration is generally about 0.1–0.2 lb/100 ft^2 of filter area or about 0.42 kg/100 m^2.

FILTER ELEMENTS AND DESIGN

Filter Performance Criteria: TIPS

What Makes a Good Filter? The characteristics of a good filter are as follows:

Fluid cleanliness or the removal of solids, which is considered as the first priority. Followed by reliability and ease of maintenance, and finally, filter life.

In a filter the removal of particles of different sizes is a function of the openings in the medium and/or the head loss across the filter.

When a filter is used to protect a piece of machinery or process, there is generally a critical size of particles, which must be removed. Filters that have high efficiency, which remove nearly all the particles that fall in the critical size ranges, will remove larger particles and help reduce maintenance costs.

All filters have a "run life," or cycle time. Length of run and removal efficiency throughout the service life of the filter are always important considerations.

Filter cloth permeability is rated on the basis of air permeability. A relatively fine mesh cloth will have a permeability of 2–3 cubic feet per minute per square foot of cloth. (600–914 liters/minute/square meter of cloth) Coarser cloths can have values above 15 scfm/sf. (4600 lpm/M^2). Filter cloths are often made from polypropylene which has an affinity for vegetable oils and greases. Filter cloths will also accumulate fine solids in their pore spaces. The oils and solids tend to reduce cloth permeability. This permeability can be restored by careful application of buffered citric and other acids and chemical cleaning with detergents. The most successful cloth cleaning is performed in the filter by recirculating the cleaning fluid. Care must be taken to prevent excessive temperature rise during recirculation and to select cleaning agents which do not attack the cloth.

Self-Cleaning Filters

Self-backwashing granular media filters are relatively new. They were first developed in the early 1980s for use in municipal wastewater treatment. The theory of developing head loss through the filter is the same as for other types of granular filters. The solids penetration in the filter is, by design, the full depth of the bed. The difference between a self-backwashing filter and

FIGURE 8.6   One view of a self cleaning sand filter.

a conventional filter is that the media bed is continually withdrawn, cleaned, and recycled to the top of the filter. In these self-cleaning filters an airlift removes the dirty particulate from the bottom of the filter bed, and subjects it to high turbulence at the entrance to the filter. The mud is decanted and dewatered for separate treatment is re-released back into the filter feed, and that clean media is returned to the filter bed. The photographs, Figures 8.6 and 8.7, show two different designs from competing manufacturers.

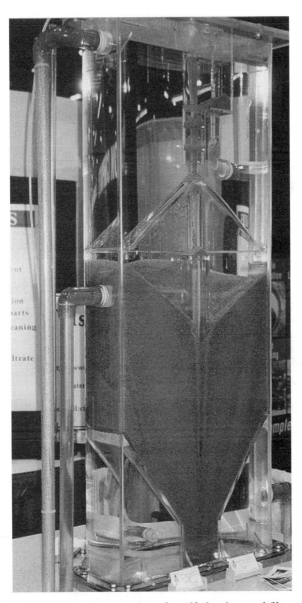

FIGURE 8.7   Cutaway view of a self cleaning sand filter.

Filter Press

The filter press is a skin filter with precoat. Two photographs of different types of presses are shown in Figures 8.8 and 8.9. Note that one is a research model, which uses steam in the chambers to provide a dryer cake for the final product.

Belt Filters

Belt filters are mechanical devices usually used on wastewater biosolids for dewatering. The material is fed into the press at the top between two belts of porous filter cloth. The belts are under tension and pass through a series of rollers as shown in Figure 8.10. The sludge is squeezed and sheared by the tension in the belts and the fact that the outer belt moving around a roller will always move faster than the inner belt, which is in contact with the roller. The belts can put the sludge under several hundred pounds of pressure per inch (100 lb/in. = 175 N/cm). The compression and shear cause the sludge to dewater to between 12% and 22% solids, depending upon the

FIGURE 8.8 Filter press with steam sterilizer tubes on top.

FIGURE 8.9   Open view of filter press.

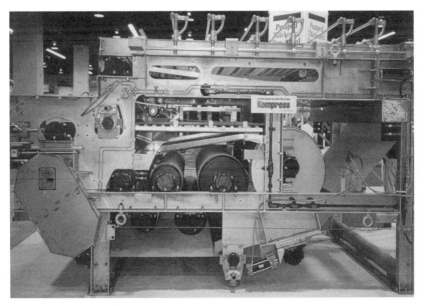

FIGURE 8.10   Belt filter press. The serpentive path around the rollers helps shear the sludge and enable it to consolidate.

feed concentration. The lower the feed, the lower the final concentration of solids. Without pretreatment, including thickening, the feed solids will not be much over 3–5%, and the effluent solids will be about 12–14%. If the feed solids are preconditioned with polymer (at high costs) and centrifiguation as pretreatment prior to the belt filter, the feed solids can attain between 10% and 15% entering and about 23% leaving. No manufacturer of belt filter presses has been able to dewater waste activated sludge to greater than about 23% solids, maximum without other amendments, such as the addition of primary sludge.

A typical belt filter press is shown in Figure 8.10.

9

DISINFECTION

General
Rate of kill—disinfection parameters
Status of U.S. drinking water
Chlorine
Ozone
Ultraviolet light
Other disinfecting compounds

GENERAL

The purpose of disinfection is the protection of the microbial water quality. The ideal disinfectant should have high bacterial toxicity, be inexpensive, and not be too dangerous to handle, and should have a reliable means of detecting the presence of a residual.

Chlorine is one of the oldest disinfection agents used, which is one of the safest and most reliable. It has extremely good properties, which conform to many of the aspects of the ideal disinfectant as mentioned above.

RATE OF KILL—DISINFECTION PARAMETERS

Chick's Law

The idea behind disinfection is to kill or to inactivate harmful bacteria and viruses.

Practical Wastewater Treatment, by David L. Russell
Copyright © 2006 John Wiley & Sons, Inc.

The time kill rate is a differential equation:

$$dN/dt = -kN$$

where k is a rate constant, and N is the number of living organisms. Note that the expression is specific to the type of organisms.

This gives $\ln(N_2/N_1) = -kt$ and $t = (2.3/k) \log(N_1/N_2)$, where the subscripts on N refer to the number of organisms at the respective times.

The rate of disinfection k is dependent upon the concentration of the disinfectant and the coefficient of dilution. The rate constant can also be affected by the temperature as shown in the Arrhenius equation:

$$k = Ce^{-(\Delta H_a/RT_a)}$$

where $\Delta H_a =$ activation energy (cal); $R =$ gas constant (1.99 cal/°C); $T_a =$ absolute temperature (K), and C is a determined constant.

The equation is evaluated by plotting **log k** versus **$1/T_a$**; factors such as nutrient concentration, pH, and osmotic pressure all affect the constants and the rate.

The death rate of microorganisms is a first-order differential equation with respect to time.

Problem: The following table shows the disinfection of poliomyelitis virus using hypobromite as a disinfectant.[1] Determine Chick's constant and the time required to reduce the concentration of viable poliovirus to 1/10,000 of the original concentration.

Viable Poliovirus Concentrations

t(s)	N/N_o
4	0.07690
8	0.00633
12	0.00050

Solution: Plot the $-\ln(N/N_o)$ against time (Fig. 9.1)

Execute linear regression for experimental points. This yields the slope of the line ($k = 0.634$/s). The time required for a 10,000-fold reduction is

$$t = [-\ln(N/N_o)]/k = -\ln(1/10{,}000)/0.634 \, \text{s} = 15 \, \text{s}$$

[1]http://www.nbif.org/course/env-engr/index.html.

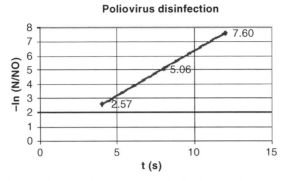

FIGURE 9.1   Sample plot of poliovirus survival ratio in disinfection experiment.

The basic organism often used in measuring disinfection efficiency is *Escherichia coli (E. coli)* but the USEPA has recently begun to focus on a number of different organisms that are more resistant than *E. coli*.

We cannot get into a discussion of disinfection without some considerations of human health factors.

For a number of years, the basic problem was *E. coli* and the principal concern was and largely still is fecal contamination of drinking water, bathing water, and so on. The *E. coli* organism was and still is the most frequent indicator of fecal contamination. However, in the past few years we have discovered that fecal streptococcus (*Streptococcus faecalis* and *S. faecium*; a subset of the fecal streptococci considered more feces specific) is a better indicator of human fecal contamination.

Giardia lamblia is a protozoan found in the feces of humans and animals that can cause severe gastrointestinal ailments. It is a common contaminant of surface waters. For a number of years, it went unnoticed because the principal focus was on coliform organisms. In 1981, the World Health Organization classified *Giardia* as a pathogen (capable of causing disease).

Physically, *Giardia* is a cyst former and can survive outside the body for long periods of time. If viable cysts are ingested, *Giardia* can cause the illness known as giardiasis, an intestinal illness, which can cause nausea, anorexia, fever, and severe diarrhea. The symptoms last for several days only and the body can naturally rid itself of the parasite in 1-2 months. However, for individuals with weakened immune systems the body often cannot rid itself of the parasite without medical treatment.[2]

---

[2]http://www.ladwp.com/bizserv/water/quality/topics/giardia/giardia.htm.

In the United States, *Giardia* is the most commonly identified pathogen in waterborne disease outbreaks—but that may be because of the attention given to *E. coli*. *Giardia* is not host-specific contrary to some forms of coliform organisms. *Giardia* can jump species, and the viable cysts excreted by animals can infect and cause illness in humans if it enters their drinking water. There are two ways in which *Giardia* can infect humans. Both involve inadequately treated (inadequately disinfected) drinking water: One way is through animal feces in the watershed entering the drinking water, and the second way is through human sewage entering the drinking water. In both the cases, the control mechanism is the adequate disinfection.

The effective control of *Giardia* is accomplished by chlorine and ozone, combined with filtration. Filtration may be sufficient by itself, but that assumes that the filtration will be sufficient to remove all the *Giardia*. The USEPA has focused on the inactivation of *Giardia* as being one key to safe drinking water.[3]

The following Web site gives more specific information on diseases and the potential problems: http://www.unc.edu/courses/envr191/191-1999.htm. The Web site has a number of specific links and good information on human pathogens, which are potential problems from waterborne diseases.

STATUS OF U.S. DRINKING WATER

The following is excerpted from a CDC report on disinfection of water systems in the United States:

Surveillance for Waterborne-Disease Outbreaks—United States, 1995–1996

by Deborah A. Levy, Ph.D., M.P.H.[1,2] Michelle S. Bens, M.P.H.[2] Gunther F. Craun, M.P.H.[3] Rebecca L. Calderon, Ph.D., M.P.H.[4] Barbara L. Herwaldt, M.D., M.P.H.[2] [1]Epidemic Intelligence Service, Epidemiology Program Office, CDC; [2]Division of Parasitic Diseases, National Center for Infectious Diseases, CDC; [3]Gunther F. Craun & Associates, Staunton, Virginia; [4]Human Studies Division, National Health and Environmental Effects Laboratory, U.S. Environmental Protection Agency.

Abstract Problem/Condition:

"Since 1971, CDC and the U.S. Environmental Protection Agency have maintained a collaborative surveillance system for collecting and periodically reporting data that relate to occurrences and causes of waterborne-disease outbreaks (WBDOs).

[3]http://www.fc.net/~tdeagan/water/one.html#GIARDIASIS.

Reporting Period Covered:

This summary includes data for January 1995 through December 1996 and previously unreported outbreaks in 1994. Description of the System: The surveillance system includes data about outbreaks associated with drinking water and recreational water. State, territorial, and local public health departments are primarily responsible for detecting and investigating WBDOs and for voluntarily reporting them to CDC on a standard form. Results: For the period 1995–1996, 13 states reported a total of 22 outbreaks associated with drinking water. These outbreaks caused an estimated total of 2,567 persons to become ill. No deaths were reported. The microbe or chemical that caused the outbreak was identified for 14 (63.6%) of the 22 outbreaks. Giardia lamblia and Shigella sonnei each caused two (9.1%) of the 22 outbreaks; Escherichia coli O157:H7, Plesiomonas shigelloides, and a small round structured virus were implicated for one outbreak (4.5%) each. One of the two outbreaks of giardiasis involved the largest number of cases, with an estimated 1,449 ill persons. Seven outbreaks (31.8% of 22) of chemical poisoning, which involved a total of 90 persons, were reported. Copper and nitrite were associated with two outbreaks (9.1% of 22) each and sodium hydroxide, chlorine, and concentrated liquid soap with one outbreak (4.5%) each. Eleven (50.0%) of the 22 outbreaks were linked to well water, eight in non-community and three in community systems.

Only three of the 10 outbreaks associated with community water systems were caused by problems at water treatment plants; the other seven resulted from problems in the water distribution systems and plumbing of individual facilities (e.g., a restaurant). Six of the seven outbreaks were associated with chemical contamination of the drinking water; the seventh outbreak was attributed to a small round structured virus. Four of the seven outbreaks occurred because of backflow or backsiphonage through a cross-connection, and two occurred because of high levels of copper that leached into water after the installation of new plumbing. For three of the four outbreaks caused by contamination from a cross-connection, an improperly installed vacuum breaker or a faulty backflow prevention device was identified; no protection against backsiphonage was found for the fourth outbreak.

Thirty-seven outbreaks from 17 states were attributed to recreational water exposure and affected an estimated 9,129 persons, including 8,449 persons in two large outbreaks of cryptosporidiosis. Twenty-two (59.5%) of these 37 were outbreaks of gastroenteritis; nine (24.3%) were outbreaks of dermatitis; and six (16.2%) were single cases of primary amebic meningoencephalitis caused by Naegleria fowleri, all of which were fatal. The etiologic agent was identified for 33 (89.2%) of the 37 outbreaks. Six (27.3%) of the 22 outbreaks of gastroenteritis were caused by *Cryptosporidium parvum* and six (27.3%) by *E. coli* O157:H7. All of the latter were associated with unchlorinated water (i.e., in lakes) or inadequately chlorinated water (i.e., in a pool). Thirteen (59.1%) of these 22 outbreaks were associated with lake water, eight (36.4%) with swimming or wading pools, and one (4.5%) with a hot spring. Of the nine outbreaks of dermatitis, seven (77.8%) were outbreaks of *Pseudomonas dermatitis* associated with hot tubs, and two (22.2%) were lake-associated outbreaks of swimmer's itch caused by *Schistosoma* species.

Interpretation:

WBDOs caused by *E. coli* O157:H7 were reported more frequently than in previous years and were associated primarily with recreational lake water. This finding suggests the need for better monitoring of water quality and identification of sources of contamination. Although protozoan parasites, especially *Cryptosporidium* and *Giardia*, were associated with fewer reported outbreaks than in previous years, they caused large outbreaks that affected a total of approximately 10,000 persons; all of the outbreaks of cryptosporidiosis were associated with recreational water, primarily swimming pools.

Prevention of pool-associated outbreaks caused by chlorine-resistant parasites (e.g., *Cryptosporidium* and to a lesser extent *Giardia*) is particularly difficult because it requires improved filtration methods as well as education of patrons about hazards associated with fecal accidents, especially in pools frequented by diaper-aged children. The proportion of reported drinking water outbreaks associated with community water systems that were attributed to problems at water treatment plants has steadily declined since 1989 (i.e., 72.7% for 1989–1990, 62.5% for 1991–1992, 57.1% for 1993–1994, and 30.0% for 1995–1996). This decrease might reflect improvements in water treatment and in operation of plants. The outbreaks attributed to contamination in the distribution system suggest that efforts should be increased to prevent cross-connections, especially by installing and monitoring backflow prevention devices.

Actions Taken: Surveillance data that identify the types of water systems, their deficiencies, and the etiologic agents associated with outbreaks are used to evaluate the adequacy of current technologies for providing safe drinking and recreational water. In addition, they are used to establish research priorities and can lead to improved water-quality regulations.

Some organisms are harder to inactivate than others. This is especially true of the spore formers and the protozoans. An example for heat disinfection is shown in Table 9.1.

The type of disinfectant is also important. The following generally holds true:

Microbe type: vegetative bacteria–viruses–protozoan cysts, spores, and eggs

　　　　　　　　least resistant - - - - - - - - - - - - most resistant

Disinfectant: O_3–ClO_2–iodine/freechlorine–chloramines

Giardia:　　best -worst

TABLE 9.1   Comparison of Bacterial Disinfection Rates

Organism	Relative Resistance
E. Coli	1
Bacterial spores	3,000,000
Mold spores	2–10
Viruses and bacteriophages	1–5

Source: O. Rahn, Physical Methods of Sterilization of Micro-organisms. Bacteriological Reviews, vol. 9, 1945, pp. 1–7.

The effectiveness of the disinfectant varies with the type of microbe and chemical and environmental factors.

Microbial aggregation: protects interior microbes from inactivation
Water quality: Particulates: protect microbes from inactivation
Dissolved organics: protects; consumes disinfectant; coats microbes.
Inorganic compounds and ions: effects vary with disinfectant
pH: effects depend on disinfectant.
Free chlorine more biocidal at low pH where HOCl predominates.
Chlorine dioxide more microbiocidal at high pH.
Free chlorine is still the most commonly used disinfectant.

Maintaining disinfectant residual during treated water storage and distribution is essential. It is a problem for O_3 and ClO_2 because they do not leave residuals and the water can be reinfected fairly easily. For these compounds, a secondary disinfectant must be used to provide a satisfactory residual. Most commonly the disinfectant chosen is chlorine.

See Figures 9.2 and 9.3 on virus and bacterial inactivation. These figures are taken from the WEF MOP #8 on Wastewater Treatment Plant Design.

CHLORINE

Silver and heat are probably the oldest disinfectants, but chlorine has got the most acceptance. Chlorine disassociates in water. The reactions are as follows:

$$Cl_2 + H_2O \rightarrow Cl_- + HOCl + H^+ \quad K_h = 4.5 \times 10^{-4}(mol/l)^2$$
$$HOCl \leftrightarrow H^+ + OCl^- \quad K_i \text{ is pH dependent}$$

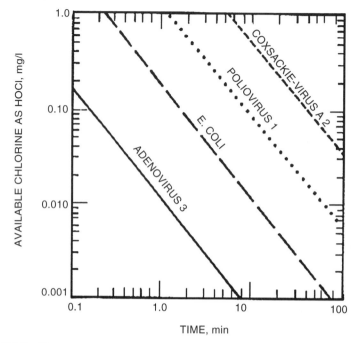

FIGURE 9.2 Time vs. concentration for 99% kill of *E. coli* and three viruses of hypochlorous acid at 0°C–6°C.

Consider the following:

The most effective form of disinfectant is the HOCl form. By applying the pH and speciating the OCl and HOCl forms, it is possible to calculate the required dose of chlorine for a specific kill based upon pH (Fig. 9.4).

Ammonia, Chlorine, and Chloramines

Free ammonia combines with the HOCl form of chlorine to form chloramines in a three-step reaction.

$$NH_3 + HOCl \rightarrow NH_2Cl + H_2O$$
$$NH_2Cl + HOCl \rightarrow NHCl_2 + H_2O$$
$$NHCL_2 + HOCl \rightarrow NCl_3 + H_2O$$

When the pH > 6 and [HOCl]/[NH₃] is around 1, monochloramine predominates. The reason for the detailed explanation is that chloramines are also a form of disinfectant—not as effective as HOCl, but as a disinfectant nonetheless.

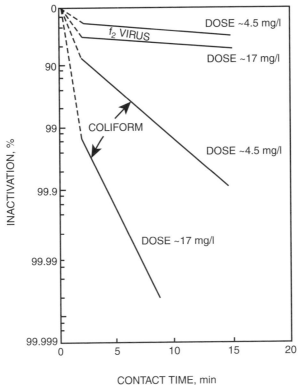

Viral and bacterial inactivation at a 5700 m³/day (1.5 mgd) conventional activated sludge plant conditions. The f_2 bacterial virus was seeded in the secondary settling basin at a titer of approximately 10^6 plaque forming units/ml. Chlorine dosages were approximately 4.5 and 17 mg/l.

FIGURE 9.3 f_2 Virus and coliform inactivation in a chlorine contact tank under controlled conditions. Viruses are often more difficult to kill than are coliform and nonspore forming organisms.

When the molar ratio of chlorine to ammonia is substantially above 2, dechlorination of the hypochlorite / hypochlorous ions occurs because of the formation of chloramines. The concentration of residual chlorine first rises then falls then rises again, as shown in Figure 9.5.

Chloramines have some disinfecting power, but their ability to inactivate viruses and especially spore formers such as *Giardia* is quite limited. Chloramines have been in use as disinfectants since the early 1900s but the

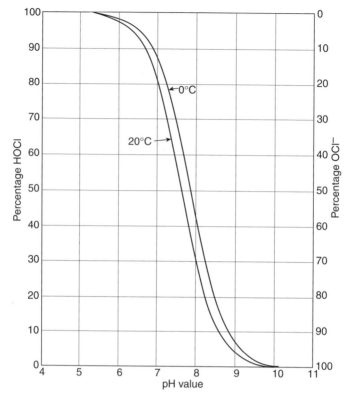

FIGURE 9.4 Distributions of hypochlorous and hypochlorite ions in water at various pH levels.

use is limited because of the expense of generation and relatively poor effectiveness against several common types of pathogens.

Other Types of Chlorine

Chlorine comes in a variety of forms. The most common are calcium hypochlorite (tablet and powder) and sodium hypochlorite (liquid). Both compounds are basic. The disassociation reactions are as follows:

$$NaOCl + H_2O \rightarrow HOCl + NaOH$$
$$Ca(OCl)_2 + H_2O \rightarrow 2HOCl + Ca(OH)_2$$

Sodium hypochlorite (concentrated liquid bleach) and calcium hypochlorite tablet and powder (dry bleach and disinfecting tablets and powders) can react violently with organics and fuels, and are corrosive to clothing.

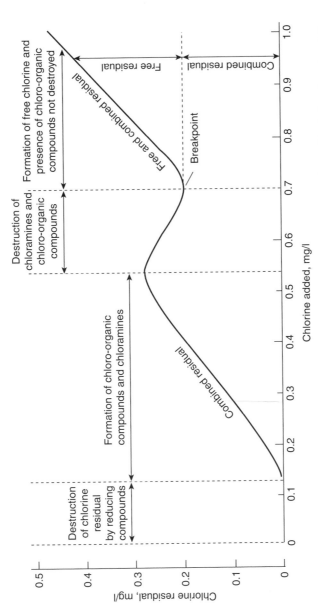

FIGURE 9.5 Break point chlorination by the formation of chloramines. The free chlorine residuals first rises then falls until the reaction with ammonia have been completed. As additional chlorine is applied and ammonia is consumed, the chlorine residual again rises.

When reacted with acids, violent explosions have been known to occur. The resulting gas is chlorine gas and hydrogen. Because it is a powerful oxidizer, it must be properly stored away from any potential fuels or reactive metals, including aluminum. There are a number of situations when people have been killed from the explosion resulting from the accidental mixing of drain cleaner [usually a sodium salt of sulfuric acid ($NaH(SO_4)$) or the acid itself (H_2SO_4)] and powdered bleach ($Ca(OCl)_2$) in trying to clean out a toilet bowl. There are an equally large number of cases of reported fires when a homeowner attempts to store oily rags in a container of calcium hypochlorite disinfecting powder, which has not been properly cleaned out and still contains powder residues.

Other Reactions with Chlorine

Chlorine in water will oxidize iron, manganese, chromium, arsenic, and a variety of other compounds. In the case of the latter two compounds the higher valence is of more toxic. It is extremely effective in oxidizing these compounds, especially at pH less than 7. It will also react with natural organic compounds such as tannins from leaves and will form trihalomethanes, chloroform, and other probable human carcinogens.[4] However, the use of the word "probable" is subject to interpretation. Various lobby groups are against the use of chlorine for a variety of reasons, and while some of the halomethanes are actual carcinogens, there is scant evidence that chlorinated drinking water will cause cancer from halomethanes, especially when the risk of not using chlorine is considered.

Chlorine forms chlorate ions that are also "suspect" compounds. The EPA is currently regulating disinfection by-products (DBP's) in municipal water supplies.

Chlorine Safety

Chlorine gas is corrosive, oxidizing, toxic, and denser than air and should be handled accordingly, with extreme caution. It can react violently with a number of compounds.

The EPA has designated chlorine as a toxic gas under Section 313 of SARA and Section 112r of the Clean Air Act. As such, anyone handling quantities in excess of 200 lb may have to fulfill special notification requirements under Section 313 and if the total quantity stored is in excess of 2500 lb of chlorine, special evacuation and community notification plans will be mandatory

---

[4]http://www.epa.gov/safewater/mdbp/mdbptg.html#disinfect.

under U.S. law. It is important to note that in the United States a one-ton cylinder of chlorine may create a "theoretical" evacuation distance around the source of almost 2 miles.[5]

Chlorine Dioxide

Chlorine dioxide (ClO_2) is a neutral compound of chlorine in the +IV oxidation state. It disinfects by oxidation; however, it does not chlorinate. It is a relatively small, volatile, and highly energetic molecule, and a free radical even while in dilute aqueous solutions. At high concentrations, it reacts violently with reducing agents. However, it is stable in dilute solution in a closed container in the absence of light.

Chlorine dioxide functions as a highly selective oxidant owing to its unique, one-electron transfer mechanism where it is reduced to chlorite (ClO_2^-).

The pK_a for the chlorite ion, chlorous acid equilibrium, is extremely low at pH 1.8. This is remarkably different from the hypochlorous acid/hypochlorite base ion pair equilibrium found near neutrality and indicates that the chlorite ion will exist as the dominant species in drinking water. The oxidation-reduction of some key reactions are as follows:

$$ClO_2(aq) + e^- = ClO_2^- \qquad E° = 0.954 \text{ V}$$

Other important half reactions are

$$ClO_2^- + 2H_2O + 4e^- = Cl^- + 4OH^- \qquad E° = 0.76 \text{ V}$$
$$ClO_3^- + H_2O + 2e^- = ClO_2^- + 2OH^- \qquad E° = 0.33 \text{ V}$$
$$ClO_3^- + 2H^+ + e^- = ClO_2 + H_2O \qquad E° = 1.152 \text{ V}$$

Chlorine dioxide is always generated on-site because it is explosive when compressed. It also reacts violently with sunlight and/or UV light with explosive decomposition. It is generated from sodium chlorite and sodium chlorate.

Chlorine dioxide is less pH sensitive than chlorine and may be equally as effective as a disinfectant. However, it costs substantially more than chlorine. Because of the air hazards associated with chlorine, and the problems associated with the treatment of various by-products of disinfection and concerns about the formation of dioxins from chlorination, chlorine dioxide has seen a

---

[5]See 40 CFR 68 for requirements and threshold requirements.

resurgence in the paper industry in the pulp bleaching area. However, it is still not as economical as chlorine.

OZONE

Ozone has the following properties:

Ozone exists as a gas at room temperature. The gas is colorless with a pungent odor readily detectable at concentrations as low as 0.02 ppm–0.05 ppm (by volume), which is below concentrations of health concern. Ozone gas is highly corrosive and toxic.

Ozone is a powerful oxidant, second only to the hydroxyl free radical, among chemicals typically used in water treatment. Therefore, it is capable of oxidizing many organic and inorganic compounds in water. These reactions with organic and inorganic compounds cause an ozone demand in the water treated, which should be satisfied during water ozonation prior to developing a measurable residual.

Ozone is slightly soluble in water. At 20°C, the solubility of 100% ozone is only 570 mg/l as compared with about 11.3 mg/l for oxygen. Typical concentrations of ozone found during drinking water treatment range from <0.1 mg/l to 1 mg/l, although higher concentrations can be attained under optimum conditions.

Ozone decomposes spontaneously during water treatment by a complex mechanism that involves the generation of hydroxyl free radicals. The hydroxyl free radicals are among the most reactive oxidizing agents in water, with reaction rates in the order of 10^{10}–10^{13} M^{-1} s^{-1}. The half-life of hydroxyl free radicals is in the order of microseconds; therefore, concentrations of hydroxyl free radicals can never reach levels above 10^{-12} M.

- In the presence of many compounds commonly encountered in water treatment, ozone decomposition forms hydroxyl free radicals. The oxidation of many types of naturally occurring organic matter leads to the formation of aldehydes, organic acids, and aldo- and ketoacids.
- Ozone can mineralize some organic materials if the pathway is predominantly one of hydroxyl radical oxidation.
- Oxidation of bromide ion leads to the formation of hypobromous acid, hypobromite ion, bromate ion, brominated organics, and bromamines.
- Bicarbonate or carbonate ions, commonly measured as alkalinity, will scavenge the hydroxyl radicals and form carbonate radicals.

Ozone Production

Because ozone is an unstable molecule, it should be generated at the point of application for use in water treatment. It is generally formed by combining an oxygen atom with an oxygen molecule (O_2):

$$3O_2 \Leftrightarrow 2O_3$$

This reaction is endothermic and requires a considerable input of energy. Ozone was first discovered by the electrolysis of sulfuric acid. Ozone can be produced by several ways, although one method, corona discharge, predominates in the ozone generation industry. Ozone can also be produced by irradiating an oxygen-containing gas with ultraviolet light and electrolytic reaction.

Corona discharge, also known as silent electrical discharge, consists of passing an oxygen-containing gas through two electrodes separated by a dielectric and a discharge gap. Voltage is applied to the electrodes, causing an electron flow through across the discharge gap. These electrons provide the energy to disassociate the oxygen molecules, leading to the formation of ozone. The following figure shows a basic ozone generator (Fig. 9.6).

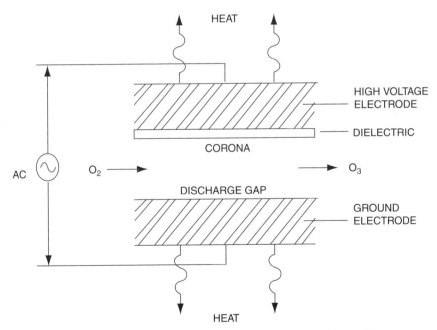

FIGURE 9.6   Schematic drawing of corona discharge method for making ozone.

ULTRAVIOLET LIGHT

Ultraviolet or UV light is a bacterial disinfectant. It carries no residual. It requires clear, un-turbid, and noncolored water for its use. Most commercial disinfection systems operate low to medium powered UV lamps and the technology currently in use focuses on a wavelength of about 354 nm (Figs. 9.7 and 9.8).

The UV dosage is calculated as

$$D = I \cdot t$$

where D = UV dose (mW $\cdot$ s/cm^2); I = Intensity (mW/cm^2); t = exposure time (s).

Research indicates that when microorganisms are exposed to UV radiation, a constant fraction of the living population is inactivated during each progressive increment in time. This dose–response relationship for germicidal effect indicates that high-intensity UV energy over a short period of time would provide the same kill as a lower intensity UV energy at a proportionally longer period of time.

The UV dose required for effective inactivation is determined by site-specific data relating to the water quality and log removal required. On the basis of first-order kinetics, the survival of microorganisms can be calculated as a function of dose and contact time.

The advantage of UV is that, for waters with high transmittance, it is directly effective against the DNA of many organisms, is not reactive with other forms of carbonaceous demand, and can give good bactericidal kill values while not leaving any residue or chlorites, or trihalomethanes.

The advantage is often the disadvantage, because power fluctuations, variations in hydraulic flow rates, and color or turbidity can cause the treatment to be ineffective. Also recently, some discussion of cell recovery and repair from UV exposure, with a consequent rapid recovery and regrowth of the damaged organisms because of the inactivation of their predators and competitors, has come to light.

OTHER DISINFECTING COMPOUNDS

Potassium Permanganate

Potassium permanganate is highly reactive under conditions found in the water industry. It will oxidize a wide variety of inorganic and organic substances. Potassium permanganate (Mn^{7+}) is reduced to manganese

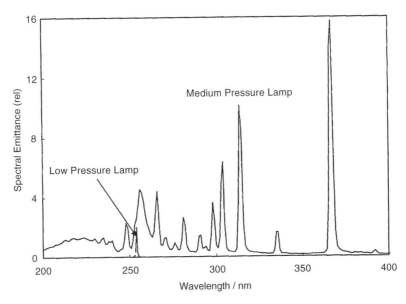

Relative spectral emittance from low pressure ($\cdots$) and medium pressure (—) lamps.

FIGURE 9.7 UV spectra for various lamps. The medium pressure lamp has the spectra almost precisely at the 357nm range where the disinfection is most effective.

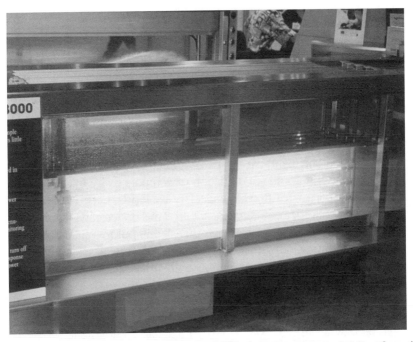

FIGURE 9.8 Horizontal lamps in a UV disinfection unit. The other potential configuration is with vertical lamps. In both cases, the flow is along the length of the lamps.

dioxide (MnO_2) (Mn^{4+}), which precipitates out of solution. All reactions are exothermic. Under acidic conditions the oxidation half-reactions are

$$MnO_4^- + 4H^+ + 3e^- \rightarrow MnO_2 + 2H_2O \qquad E° = 1.68\,V$$
$$MnO_4^- + 8H^+ + 5e^- \rightarrow Mn^{2+} + 4H_2O \qquad E° = 1.51\,V$$

Under alkaline conditions, the half-reaction is

$$MnO_4^- + 2H_2O + 3e^- \rightarrow MnO_2 + 4OH^- \qquad E° = 0.60\,V$$

Reaction rates for the oxidation of constituents found in natural waters are relatively fast and depend on temperature, pH, and dosage.

Potassium permanganate is a good oxidant but not a very good disinfectant. Its primary uses are taste and odor control, and like HOCl it is more effective as a disinfectant under acidic conditions down to a pH of about 5.9. Under alkaline conditions, it is very powerful as an oxidant but less so as a disinfectant.

Hydrogen Peroxide and Ozone

Hydrogen peroxide is a liquid with the formulation H_2O_2. There is a bit of a debate over the exact mechanism of disinfection. Hydrogen peroxide is not much of a disinfectant by itself, but in combination with ozone it has powerful disinfection properties. The combination is called peroxone. Until recently there was a large debate on whether or not peroxones even existed. One school of thought discusses the formation of peroxones, another school of thought states that the peroxones do not exist.[6] Hydrogen peroxide or ultraviolet radiation accelerates the decomposition of ozone and increases the hydroxyl radical concentration. By adding hydrogen peroxide, the net production of hydroxyl free radicals is 1.0 mole hydroxyl radical per mole ozone. The two principal methods of disinfection are (1) direct oxidation of compounds by aqueous ozone ($O_{3(aq)}$) and (2) oxidation of compounds by hydroxyl radicals produced by the decomposition of ozone. Dosage levels for peroxide and ozone are generally in the order of 5 mg/l each with ratios of peroxide/ozone between 0.5 and 0.8 and detention times greater than 5 min but less than 20 min.

The two oxidation reactions compete for substrate (i.e., compounds to oxidize). The ratio of direct oxidation with molecular ozone is relatively slow (10^{-5}–$10^7\,M^{-1}\,s^{-1}$) compared with hydroxyl radical oxidation

---

[6]A paper by Xin Xu and William A Goddard III published in the November 2002, Proc. National Academy of Sciences, Vol. 99, No. 24 has done much to identify the complex formation of peroxones—see "Peroxone Chemistry: Formation of H_2O_3 and ring (HO_2)(HO_3) from O_3/H_2O_2."

($10^{12}-10^{14}$ M^{-1} s^{-1}). The hydroxyl radical reactions are very fast, but the concentration of hydroxyl radicals under normal ozonation conditions is relatively small.

A key difference between the ozone and peroxone processes is that the ozone process relies heavily on the direct oxidation of aqueous ozone while peroxone relies primarily on oxidation with hydroxyl radical, which is a powerful oxidant in its own right. In the peroxone process, the ozone residual is short-lived because the added peroxide greatly accelerates the ozone decomposition. The oxidation by the hydroxyl radical more than compensates for the reduction in direct ozone oxidation because the hydroxyl radical is much more reactive. The net result is that oxidation is more reactive and much faster in the peroxone process compared with the ozone molecular process. However, because an ozone residual is required for determining disinfection CT credit, peroxone may not be appropriate as a predisinfectant.

Because the ozone peroxide radical oxidation is much more vigorous and effective than with ozone oxidation alone, it is being used to treat organics, which are difficult to oxidize, such as taste and odor compounds and chlorinated organics (PERC and TCE) and reactive materials including explosives in the groundwater.

Neither ozone nor peroxone significantly destroys TOC. Peroxone will oxidize the saturated hydrocarbons and produce by-products such as aldehydes, ketones, peroxides, bromate ion, and biodegradable organics. However, because the peroxone is a "more powerful and rigorous oxidant," the organic material is subsequently rendered more amenable to hydrolysis and subsequent oxidation by bacterial compounds and can be biodegraded.

pH and bicarbonate alkalinity play a major role in peroxone effectiveness because the carbonate/bicarbonate system competes for hydroxyl radical at high alkalinity and at high pH levels. The presence of fine particulate solids causing turbidity does not affect the effectiveness of peroxone treatment, and the presence of peroxones will not necessarily reduce turbidity.

Table 9.2 summarizes the key differences between ozone and peroxone as they relate to their application in drinking water treatment.[7] The comparisons are similar for wastewater treatment.

Bromine and Iodine

Bromine Bromine has been used as a disinfectant in a number of applications. It has good toxicity, is a liquid at room temperatures (while chlorine is a gas at room temperature) is somewhat easier to handle than chlorine.

---

[7]Alternative Disinfectants and Oxidants Guidance Manual EPA 815-R-99-014, April 1999.

TABLE 9.2   Comparison Between Ozone and Peroxone Oxidation

Process	Ozone	Peroxone
Ozone decomposition rate	"Normal" decomposition producing hydroxyl radical as an intermediate product	Accelerated ozone decomposition increases the hydroxyl radical concentration above that of ozone alone
Ozone residual	5–10 min	Very short-lived owing rapid reaction
Oxidation path	Usually direct aqueous molecular ozone oxidation	Primarily hydroxyl radical oxidation
Ability to oxidize iron and manganese	Excellent	Less effective
Ability to oxidize taste and odor compounds	Variable	Good, hydroxyl radical more reactive than ozone
Ability to oxidize chlorinated organics	Poor	Good, hydroxyl radical more reactive than ozone
Disinfection ability	Excellent	Good, but systems can only receive CT credit if they have a measurable ozone residual
Ability to detect residual for disinfection monitoring	Good	Poor, cannot calculate CT value for disinfection credit

The chemistry of bromine is similar in many respects to the chemistry of chlorine; however, bromine cannot be used for shock treating (high dose disinfection) in the same manner as chlorine is used.

Bromine has a pH of 4.0–4.5. When bromine is added to water and an oxidizer is present, the bromine forms hypobromous acid (HOBr) and hypobromite ions (OBr). Like chlorine, the percentage of each is affected by pH. However, the effect is not as strong as it is with chlorine. Like chlorine, bromine combines with organic impurities to form combined bromine or bromamines. However, combined bromine is still an effective sanitizer, and it does not smell as strongly as chlorine. Bromine is substantially more expensive than chlorine. Consequently, it has fallen out of use as a commercial disinfectant except in swimming pools where it is still used, because it reportedly has less eye irritation than chlorine, but it has seen a resurgence in popularity because of the perceived hazards associated with chlorine gas. Bromine's disinfectant power is also dependent upon pH as shown in Table 9.3.

TABLE 9.3   Ion Species of Bromine with pH (Compare to Fig. 9.4 for Chlorine)

HOBr Hypobromous Acid % Bromine as HOBr	pH	OBr⁻ Hypobromite Ion % Bromine as OBr⁻
100.0	6.0	0.0
99.4	6.5	0.6
98.0	7.0	2.0
94.0	7.5	6.0
83.0	8.0	17.0
57.0	8.5	43.0

Iodine[8]   Iodine kills bacteria and disease-causing organisms. Iodine is, however, ineffective as an algicide. Iodine has been in use to disinfect water since the early 1900s. In its natural state, iodine is a solid black crystal. The simplest method of disinfecting water with iodine is by dissolving iodine in water to form a saturated solution and then injecting the iodine solution into a water system.

Iodine does not kill bacteria on contact; a holding time of at least 20 min is needed depending on the iodine concentration. An iodine residual of 0.5–1.0 mg/l should be maintained, and iodine at this level gives the water little or no iodide taste or odor. Iodine can be removed from water with a carbon filter just before drinking.

Iodine dosage is highly temperature dependent—iodine crystals are more soluble at higher temperatures. Iodine remains effective over a wide range of pH and does not lose effectiveness until the pH of water reaches 10. Iodine residuals in water can easily be measured using a test kit that indicates a color change.

Iodine tablets were developed during World War II to disinfect small amounts of water for emergency or temporary use. A few drops of tincture of iodine or iodine tablets are popular with campers and the military for disinfecting water.

Types of Iodinators

Iodine solutions are injected into a water system using bypass saturator systems or injection pumps. A holding tank or coil of pipe is used after iodine injection to provide the necessary holding time.

[8]http://www.ag.ohio-state.edu/~ohioline/b795/b795_10.html.

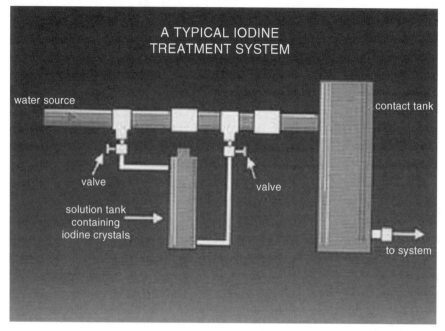

FIGURE 9.9 Schematic drawing of a bypass iodinator—United States patent 4555347.

The most common type of iodinator is called a bypass saturator and consists of a solution tank containing iodine crystals. Bypass saturators do not require any electrical connections. The solution tank is connected to the water system and diverts a small amount of water through it and back into the water line. Valves are placed on either side of the iodinator to control the iodine dose. Fluctuation in water temperature affects the solubility of iodine. Adjustments in the bypass rate are needed if water temperature changes.

Chemical injection pumps can also be used to inject iodine solutions for individual water treatment. These are the same injection systems that are used for chlorine.

Iodinators are in-line systems that are sized to treat all the water used in a household (Fig. 9.9).

Careful Use of Iodine

The question of possible health effects of iodine is still unanswered. No adverse health effects have been shown, yet continuous consumption of iodine-treated water is not recommended. Carbon filters can be used to remove iodine just before drinking. Iodine is also appropriate for occasional use in vacation homes, campgrounds, and restaurants. Iodine treatment of drinking water supplies to dairy cattle is also a concern. Because dairy cattle can drink from 15 gallons to 30 gallons of water a day, and normal levels of iodine used for disinfection may cause iodine carryover into milk.

10

NITROGEN REMOVAL

NITROGEN CHEMISTRY AND FORMS

The principal forms of nitrogen are ammonia, nitrite, and nitrate. Occasionally, one encounters N_2O, but that is infrequent. Organic nitrogen is often found in wastewater, but it is generally tied up with biosolids and is removed through precipitation or sedimentation.

The first bit of information necessary is the understanding of how the nitrogen cycle works and how the various types of nitrogen are related (Fig. 10.1).

The principal reactions associated with ammonia to nitrate pathway are the following:

$$2NH_4^+ + 3O_3 \rightarrow 2NO_2^- + 4H^+ + 2H_2O$$
$$2NO_2^- + O_2 \rightarrow 2NO_3^-$$

The first reaction takes place with nitrosomonas. The second reaction takes place with nitrobacter. However, the rate of reaction of nitrobacter is

Practical Wastewater Treatment, by David L. Russell
Copyright © 2006 John Wiley & Sons, Inc.

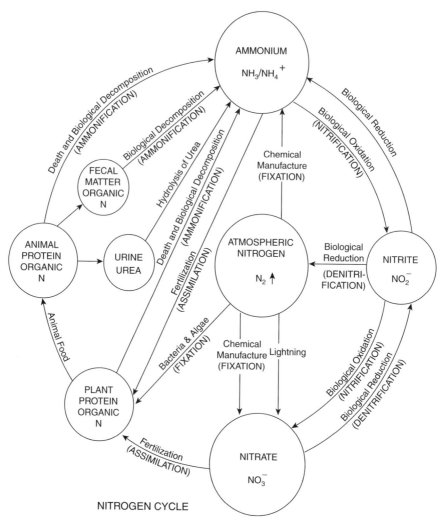

FIGURE 10.1 The nitrogen cycle.

about three times the reaction rate of nitrosomonas, and nitrite does not accumulate.

The overall reaction is that it takes about 4.6 mg/l of oxygen to oxidize 1 mg/l of ammonia completely.

Denitrification

Denitrification is accomplished by a number of bacteria—*Psuedomonas, Micrococcus, Achromobacter*, and *Bacillus*.

The principal reactions for denitrification are as follows:

$$NO_3^- + 0.33CH_3OH \rightarrow NO_2 + 0.33CO_2 + 0.67H_2O$$
$$\text{(carbon source)}$$

and

$$NO_2^- + 0.5CH_3OH \rightarrow 0.5NO_2 + 0.5H_2O + 0.5CO_2$$

where methanol is the principal carbon source for the reaction, but not necessarily the only carbon source to make the reaction proceed. We will come back to this later.

AMMONIA TOXICITY AND NITROGEN LOADING

Ammonia

The chemistry of ammonia is relatively simple and straightforward, especially in aquatic systems. Ammonia exists in two forms, the ammonium ion (NH_4^+) and un-ionized ammonia (NH_3). Organic nitrogen may contain ammonia as well, but it is generally tied up in the organic molecule and not available as a radical.

Un-ionized ammonia is highly toxic to some forms of aquatic life, while the ammonium ion is significantly less toxic. The distribution of ammonia between those two forms in water is controlled by pH, temperature, and ionic strength. In freshwater sediments at pH 8.0 and 27°C, about 3% of the total ammonia is present in the un-ionized form, while in saline water of the same temperature and pH, about 4% of the total ammonia is in un-ionized form.

The USEPA's chronic water quality criterion for un-ionized ammonia in marine waters is 0.035 mg/l NH_3 based on the sensitivity of various types of shrimp and fish. The comparable saltwater final acute value for ammonia is 0.465 mg/l NH_3. For freshwater, the USEPA has established an unionized ammonia 4-day average chronic standard of about 0.02 mg/l NH_3. When the pH decreases and the temperature decreases, the standard value falls to approximately 0.01 mg/l NH_3. Acute:chronic ratios range from 3 mg/l to 43 mg/l NH_3; 96-h LC_{50}s were reported as low as 0.08 mg/l NH_3. The results vary widely depending upon the species used to run the test, and salmonids are among the most sensitive to ammonia.

In short, ammonia, especially un-ionized ammonia (anything over pH about 8.3) is toxic to many forms of aquatic life. Ammonia complexes with other metals can also increase toxicity, especially some of the heavier metals such as nickel, cadmium, and so forth. Many biological treatment plants, except those that operate with an extremely long sludge retention time, such as extended aeration, cannot successfully meet the ammonia standard in the effluent, and ammonia toxicity has become a problem in a number of municipal and industrial wastewater treatment plants.

NITRATE

The prime concern in nitrate is water quality, and specifically drinking water quality. Many bacteria and algae have the ability to fix nitrogen from the air. Depending upon the overall ability of a stream or watershed to treat various types of pollution and the categorical usage of the stream (recreation, sport fishing, swimming, and so on—see Chapter 1 on Water Quality) one may find a total nitrogen load (along with a phosphorous load) and/or an ammonia limit on the water quality of the stream. Nitrate is generally nondeleterious and provides oxygen in times of biological stress on the river.

NITROGEN REMOVALS

There are several methods of removing nitrogen. The first is nitrification followed by denitrification. The second is ammonia stripping. The third is weak ion exchange. We will be looking at all of them.

Nitrification

Ammonia is often a by-product of incomplete treatment such as in activated sludge and contact stabilization.

Ammonia can be oxidized to nitrate nitrogen by one of the two principal routes. The reactions given in the section 'Nitrogen Chemistry and Forms' are in brief. In reality, both nitrobacter and nitrosomonas reactions are somewhat more complex because the bacteria also produce solids in the form of new cells and also respire. Both these processes consume energy. The overall reactions shown below yield 0.15 mg/cells per mg of $NH_4^+ - N$ destroyed and 0.02 mg/cells per mg of NO_2. It is also important to note from the following equation that nitrification also destroys a lot of alkalinity in the water.

The overall combined reactions are shown in the following equations:

$$55NH_4^+ + 76O_2 + 109HCO_3^- \rightarrow C_5H_7NO_2 + 54NO_2^- + 57H_2O + 104H_2CO_3$$
Nitrosomonas

$$400NO_2^- + NH_4^+ + 4H_2CO_3 + HCO_3^- + 195O_2 \rightarrow C_5H_7NO_2 + 3H_2O + 400NO_3^-$$
Nitrobacter

The overall reaction is:

$$NH_4^+ + 1.83\,O_2 + 1.98\,HCO_3^- \rightarrow 0.021\,C_5H_7NO_2 + 1.041\,H_2O$$

What is known is that while the theoretical value of 7.1 mg alkalinity is destroyed for every mg of NH_4 oxidized, in practice the actual measured values are between 6.3 and 7.4 mg alkalinity destroyed/mole of NH_4 oxidized.

During the process of nitrification, the pH of the liquid may be affected because of the destruction of alkalinity.

$$pH = pK_1 - \log(H_2CO_3)/(HCO_3)$$

For example, in a system where there is 20 mg/l of NH_3 nitrified, it will destroy about 143 mg/l of alkalinity, if there is sufficient alkalinity, or it just might depress the pH and stop the reactions.

Now the kinetic constants for growth of the nitrifying bacteria are as follows:

$$K_N = 10^{(0.51T-1.158)}$$

where T is the temperature (°C).

This is the half-saturation constant for oxidation of ammonia nitrogen.

Temperature also has an effect on *Nitrosomonas*. The temperature effect is shown in Figures 10.2 and 10.3. The temperature affects both the half-saturation constant and the overall growth rate.

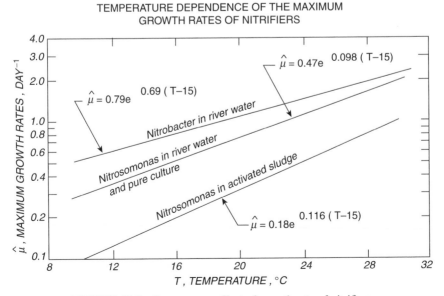

FIGURE 10.2   Temperature effect of growth rate of nitrifiers.

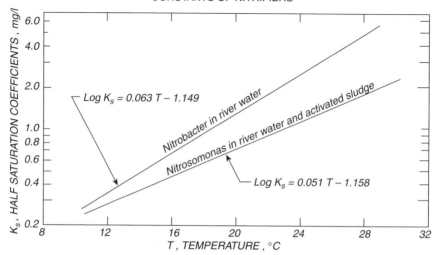

FIGURE 10.3   Temperature effect of half-saturation constant on nitrifiers.

There is some indication that the completeness of the nitrification reaction is also dependent upon maintaining a minimum oxygen level of 2 mg/l in the aeration basin, and preferably 4 mg/l because the oxidation is more complete. This can be related to the biological half-reaction rates, but it is easier to just keep the DO at between 2 and 4 mg/l.

The data on the following table are taken from a USEPA manual on "Process Design for Nitrogen Control." This manual, though older, still presents and represents some of the current technology, and it contains a complete discussion on design parameters. It is recommended for further examination, as many of the figures used in this chapter came from that source (Fig. 10.4).

In summary, for good nitrification and low ammonia effluent values, a high sludge age (biological solids retention time in the aeration basin) is preferred. That suggests that the MLSS in the aeration chamber should be as high as practical and/or that the system should be designed around extended aeration. Sludge ages beyond 10 days up to 30 days give good conversion of ammonia. Sludge ages over 30 days are considered excessive for other reasons.

Excess ammonia in the influent to a wastewater treatment plant causes excessive oxygen demand as well as the need to re-balance the carbon: nitrogen:phosphorous ratio in the wastewater, for optimum biological growth. One way of solving this problem led to a system of treatment in

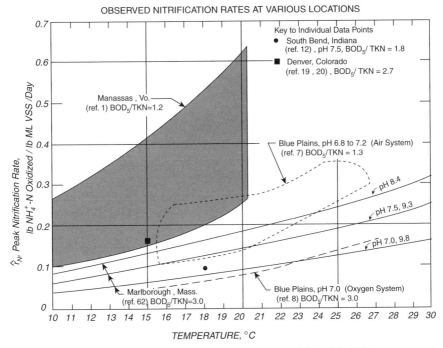

FIGURE 10.4 Various nitrification rates as published by EPA.

which the ammonia was removed first, before treatment. This avoids the need for a supplemental carbon source added after the main aeration chamber.[1]

In the presence of a carbon source such as raw wastes, and zero oxygen, and the correct supply of other nutrients in the wastewater, and lots of live bacteria, the bacteria will continue to respire. If no oxygen is supplied to replenish the depleted oxygen the bacteria will continue to respire and reduce the oxygen levels to zero (anoxic). At that point the bacteria continue to respire only using nitrate as an oxygen source. The reactions are quite complex involving the adenozine diphosphate (ADP) and adenozine triphosphate (ATP) chains and involve a number of bacteria. The bacteria that can reduce nitrate to nitrogen include *Achrombacter, Bacillus, Aerobacter, Micrococcus, Alcaligenes, Flavobacteria*, and *Proteus*. They are all facultative and fit into the general category of chemoorganotrophs.[2] The general stoichiometric equation for denitrification is: $NO_3^- + 1.2\,H^+ + 5\,e^- \Rightarrow 0.5\,N_2 + 3\,H_2O$. When COD is used as a carbon source, the production of biomass from

---

[1]Almost any carbon source can be utilized. Simple sugars have been utilized as well.

[2]Derin Orhon and Nazik Artan, Modeling of Activated Sludge Systems, Lancaster, PA: Technomic Publishers, 1994, p. 398.

	Substrate Level Denitrification	Endogenous Level Denitrification	Separate Stage Nitrification	Mixed Nitrification/Denitrification	Fixed Film	Chemical (methanol) Addition	Internal Recycle
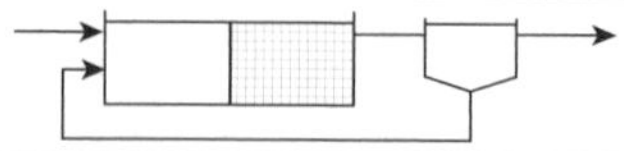 1. Wuhrman		✗					
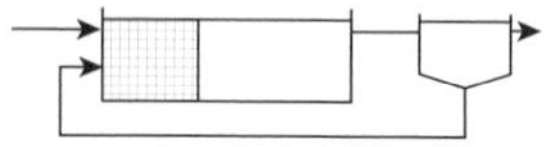 2. Ludzack-Ettinger	✗						
3. MLE	✗					✗	✗
4. Bardenpho	✗	✗				✗	✗
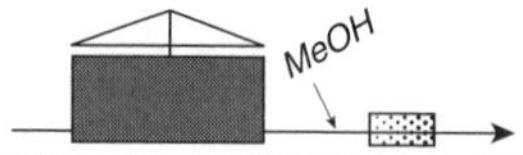 5. Trickling Filter Filter			✗		✗	✗	
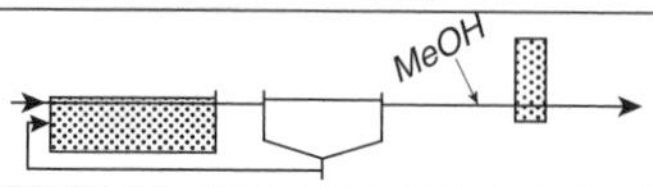 6. Activated Sludge Fluidized Bed			✗		✗	✗	
7. SBR				✗			
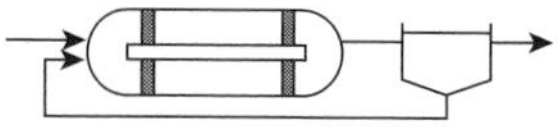 8. Oxidation Ditch				✗			
9. Biodenitro				✗			
10. Biolac				✗			
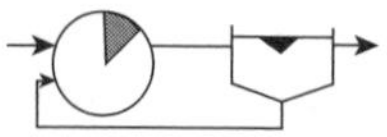 11. Counter Current Aeration				✗			
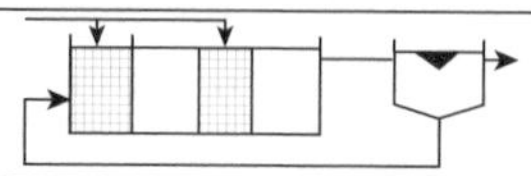 12. Step Feed Denitrification	✗	✗					

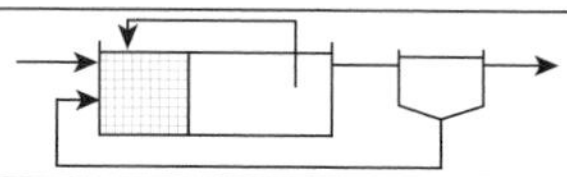

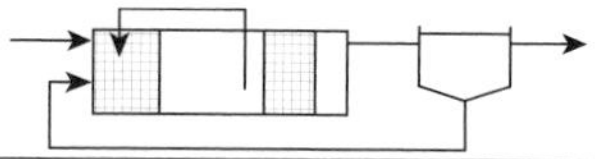

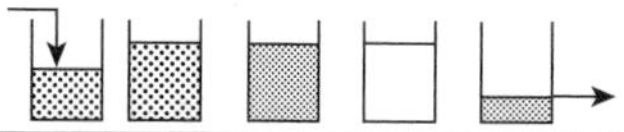

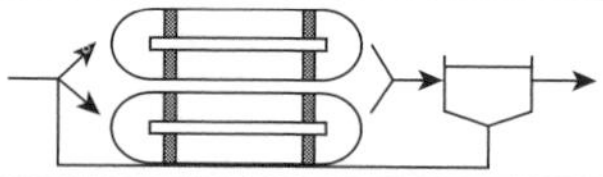

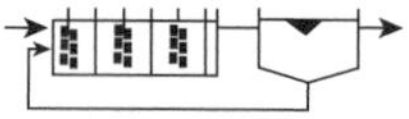

Typical biological nitrogen removal schemes.

FIGURE 10.5 System for nitrogen removal.

SEQUENTIAL CARBON OXIDATION-NITRIFICATION-DENITRIFICATION

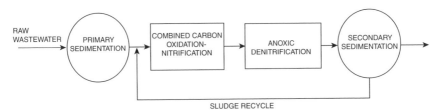

FIGURE 10.6 Nitrification/denitrification system for nitrogen removal. The process control is somewhat difficult because the wastewaters can become strongly anaerobic in the clarifier and generate gas bubbles, defeating the purpose of the clarifier, also the wastewater will have to be aerated prior to discharge in order to comply with stream standards.

anoxic synthesis is lower than when methanol is used as a carbon source. The respective values for methanol and COD are 0.55 g cells produced per gram methanol consumed and 0.30–0.25 g cells per gram COD consumed.

Several systems were developed to reduce the nitrogen in the system before it got into the aeration tank. The most common manner of removal was to turn off the oxygen supply to the head end of the wastewater treatment plant where the return sludge (rich in "hungry" bacteria from the clarifier) was mixed with the incoming waste feed. This effectively turns one end of the aeration system (some designs use separate tanks) into an anoxic zone, and the bacteria in the system would be starved for oxygen and would turn to the nitrogen compounds to reduce nitrate and ammonia to gaseous nitrogen.

There are a number of processes for nitrogen removal shown in Figures 10.5–10.7. One of the most popular innovations is the BardenPho process, which we will use for further study:

THE BARDENPHO SYSTEM - SEQUENTIAL UTILIZATION
OF WASTEWATER CARBON AND ENDOGENOUS CARBON

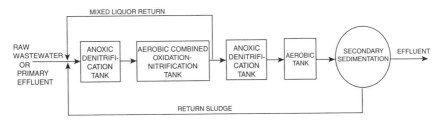

FIGURE 10.7 The Bardenpho system for nitrogen removal can also be used for phosphorous removal with only slight modification.

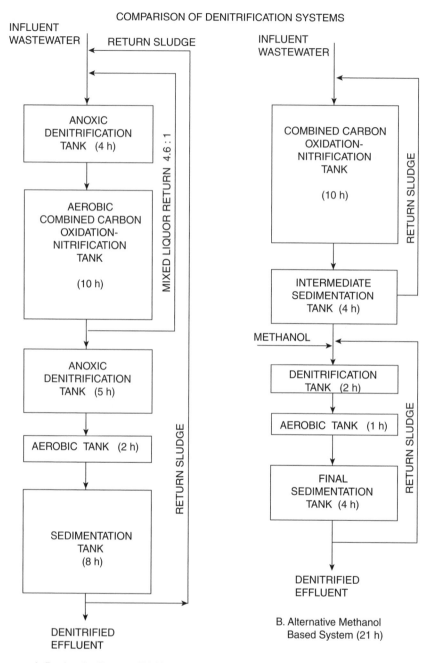

COMPARISON OF DENITRIFICATION SYSTEMS

A. Bardenpho Process (29 h)

B. Alternative Methanol Based System (21 h)

FIGURE 10.8   The advantage of using methanol as a carbon source for nitrogen removal results in substantially smaller tank sizes and capital investment.

TABLE 10.1   Compounds Toxic to Nitrifiers

Organics	Inorganics
Thiourea	Zn
Allyl-thiourea	OCN^-
8-hydroxyquinoline	CIO_4^-
Salicyladoxine	Cu
Histidine	Hg
Amino acids	Cr
Mercaptobenzthiazole	Ni
Perchloroethylene	Ag
Trichloroethylene	
Abietec acid	

HA Painter, "Review of literature on Inorganic Nitrogen Metablolism," Water Research No. 4, No. 6. pp. 393–450.

A comparison of the two types of systems (methanol versus BardenPho) is shown in Figure 10.8:

Cautionary note: There are a number of things in industrial wastes that are toxic to nitrifiers (Table 10.1).

Overall the BardenPho system is very economical when compared with the cost of operation. One pays a penalty for tank sizing and equipment, but that is amortized relatively quickly when the cost of methanol is considered for a carbon source. Methanol is an expensive source for carbon in the process, and research has shown that other sources of carbon such as sugars can be substituted at little or no penalty, and a substantial savings in cost.

Ammonia Stripping

Figure 10.9 illustrates that ammonia ionization is pH dependent.

To remove ammonia completely, raise the pH to 11.5 and blow air through the wastewater. The effectiveness of the tests has been confirmed at the Blue Plains Treatment Plant in Washington, DC. Of course the problem with this is that all wastewater treatment plants have a discharge pH limit between 6.0 and 9.0, and the fact that the high pH is inhibitory to the biological activity in the plant, so the wastewater must be neutralized back to a lower pH where biological life is encouraged.

As with any separation process, the significant variables include packing types and depth of packing, as shown in Figure 10.10:

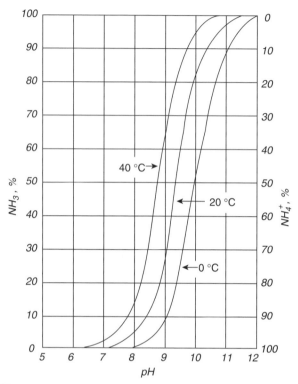

EFFECTS OF pH AND TEMPERATURE ON DISTRIBUTION
OF AMMONIA AND AMMONIUM ION IN WATER

FIGURE 10.9   Partitioning of ammonia owing to temperature and pH.

There is some indication that packing size and shape do make a substantial difference.[3] Scale from the aeration of highly alkaline wastewater has caused plugging and packing failure due to carbonate buildup. This can be avoided by paying some attention to design (Fig. 10.11).

The challenges, with stripping are, as always, slime controls, temperature, and power. Wintertime operations can inhibit removals of ammonia due to lower temperatures. Use of a tower for removal of ammonia will also dramatically cool the water and could affect the remaining biological

---

[3]For air flows of around 1000 ft³/gallon, in a 24 ft tower, a comparison of 1.5″ × 2″ redwood slats was made against 4″ × 4″ plastic truss bars at South Lake Tahoe WWTP, and the redwood slats showed virtually 100% ammonia removal versus 75% for the truss bars. The truss bars never did reached the same level of effectiveness that the slats attained even for air flows up to 4000, the ammonia removal was only 90%. The reasons are somewhat obvious—surface area being a significant factor. See Slechta, Culp. JWPCF 1967, Vol. 39, No 5., pp. 787–814.

PERCENT AMMONIA REMOVAL VS. SURFACE LOADING RATE
FOR VARIOUS DEPTHS OF PACKING (REFERENCE 10)

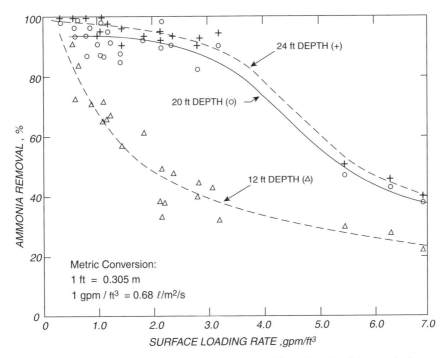

FIGURE 10.10   Effectiveness of ammonia stripping with packing depth in a packed tower.

processes. In addition, there is always the challenge of the tower getting freezed in cold weather. Although this is often a pretty sight, it represents an operational headache that should be avoided.

Ion Exchange

A number of naturally occurring "weak ion exchange media" (zeolites) have been used to remove ammonia. One of the most common is clinoptilolite, which is described as follows:

A hydrated alkali aluminosilicate that is one of the most abundant minerals in the zeolite family. Its structure consists of an outer framework of silica and alumina tetrahedra, within which water molecules and exchangeable cations (e.g., calcium, potassium, sodium) migrate freely. Although clinoptilolite's chemical formula varies with composition, a typical representation is given by $(Na_2, K_2, Ca)_3Al_6Si_{30}O_{72}24H_2O$.

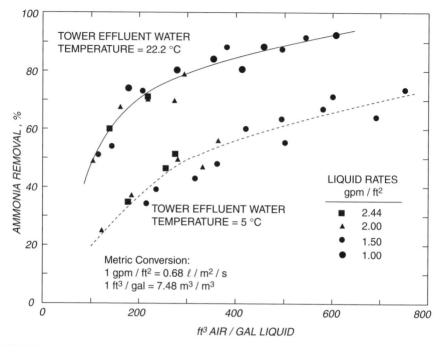

EFFECT OF TEMPERATURE ON AMMONIA REMOVAL EFFICIENCY
OBSERVED AT BLUE PLAINS PILOT PLANT(REFERENCE 9)

FIGURE 10.11   Effect of temperature on ammonia stripping at Blue Plains Wastewater Treatment Plant (POTW) in Washington, DC.

Clinoptilolite's structure closely resembles that of heulandite, another zeolite mineral, but contains a higher proportion of silica and alkalies. Clinoptilolite is somewhat soft and forms platy, nearly transparent crystals of monoclinic symmetry. It is typically colorless in thin sections, but other colors (e.g., brown, pink, red) may occur owing to the presence of impurities such as iron oxide. The dehydrated mineral has the properties of a molecular sieve that selectively extracts nitrogen from a stream of air, leaving the effluent enriched in oxygen. As an ion exchanger, clinoptilolite has been used to remove cesium and strontium from radioactive wastes produced in reprocessing nuclear fuels and to remove ammonia from sewage streams. The mineral is also used as a filler and bulking agent in the manufacture of paper.

Clinoptilolite can be found in many zeolitic sedimentary rocks; in the compacted deposits of volcanic ash commonly called tuffs; as a byproduct of the weathering of basalt; and in some shale deposits. Its sites of occurrence include Oregon, South Dakota, and Wyoming, U.S.; New Zealand; New South Wales, Australia; the Faroe Islands; and Bombay, India.

— *Source*: Britannica.

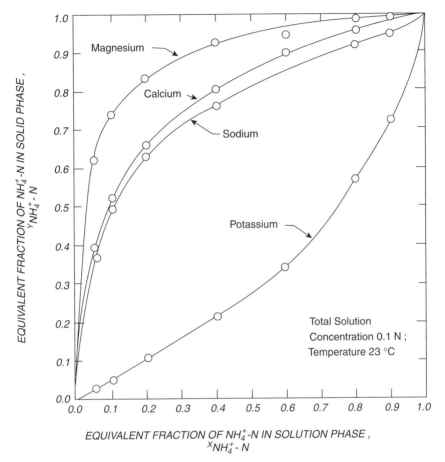

FIGURE 10.12   Isotherms for ammonia removal via ion exchange on clinoptilolite.

These materials will be discussed more broadly in Chapter 16, but observe the following (Figs. 10.12 and 10.13):

Other exchange media are available as well, but are significantly more expensive than the natural rock materials. The difference in the cost of the exchange media can be substantial. The problems commonly encountered with the backwash of the zeolites are well identified in the literature. Many facilities use a 2% sodium solution to backwash the zeolites at a neutral pH. The ammonia can then be stripped or recovered by other processes.

The problem with backwashing an ion exchange process is the following: Assume that you have an ion exchange process where you are removing

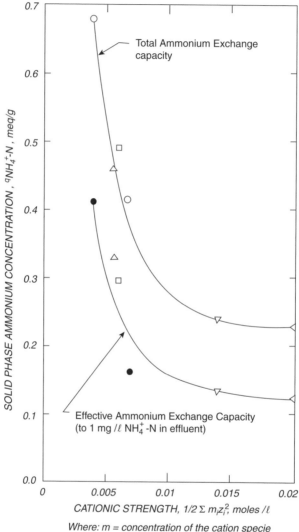

FIGURE 10.13   Ammonium ion exchange capacity in competition with other ions in wastewater on clinoptilolite.

ammonia. For a typical process, you might have a concentration factor on the basis of between 10:1 and 40:1 for backwash the volume throughput. When you have the regenerated solution, it will be a 2% brine contaminated with high levels of ammonia and all the other ions you have removed from the waste stream. The process is nondiscriminating and will not necessarily remove only ammonia cations but will remove other metals as well, depending upon their ionic strength and valence. It is conceivable that if there are metals as well in the waste stream, one could have a hazardous waste liquid, and what are you to do with that? The question is where do you go with the brine, because even if it does not have hazardous metals in it, it still must be treated for disposal.

MIXED MEDIA AND ATTACHED GROWTH SYSTEMS

Some of the wastewater treatment plants have designed mixed media upflow and downflow reactors for ammonia removal. These systems have a higher rate because the bacteria attach themselves to the media, and the media concentration is somewhat denser than in a suspended growth system unless the mixed liquor is kept very high.

These reactors can have dumped packing, and graded packing (sand granular mixed media—including gravel and rocks), and a variety of other mechanisms to create surface area for attached growth. The kinetics for this type of denitrification are essentially the same as the kinetics for suspended growth systems. A detailed discussion of the technology and operation can be found in the *Journal of the Water Environment Federation* (*Journal of the Water Pollution Control Federation*) from the late 1970s through the mid-1980s.

CONCLUSIONS

Nitrogen removal is somewhat of a pay-me now or pay-me later choice. If one uses extended aeration systems and manipulates them in ways to generate a highly nitrified effluent, additional stirred tankage (anoxic versus anaerobic) with activated sludge will provide nitrate removal. The key lies in the correct establishment of the reaction rate kinetics. For industrial wastes that will be waste-specific. The reaction kinetics must be well understood and preferably tested before the plant is built.

Ammonia removal is also the same type of choice, a balancing act between addition of a carbon source for nitrogen removal, or investment of additional capital for new tankage and operation of the system to maximize nitrogen removal.

There is some indication from research that the use of methanol may be beneficial if the removal rates for nitrogen are to be higher. There may be a limiting rate for the process without a readily attainable carbon source.

11

PHOSPHOROUS REMOVAL

General
Biological phosphorous removal
Chemical phosphorous removal

GENERAL

Phosphate in effluents from a wastewater treatment plant is responsible for many growths of unicellular blue-green algae and many other types of algae. Phosphate has been determined to be the principal limiting nutrient in controlling algae growths. The most prevalent source of phosphates was detergents, but feedlot operations (cattle and chicken) and excessive agricultural application of fertilizers are also major contributors; the wastewater treatment plants are point sources and tend to be more easily controlled.[1] The feedlot operations are often regulated as point sources, but the agricultural fertilizer applications are considered an area source and are much more difficult to regulate. The presence of natural deposits of phosphates in some rock formations (parts of Tennessee and Florida) can also contribute to the overall phosphate loading for a stream but the phosphate is generally in the calcium phosphate form and is, to a lesser extent, bioavailable.

According to a U.S. Geological Survey water quality study report in 1999, the concentration of phosphates in pristine rivers is less than 0.02 mg/l, and

---

[1]After many states instituted phosphate bans, the detergent industry voluntarily eliminated phosphate as a detergent booster in 1994.

Practical Wastewater Treatment, by David L. Russell
Copyright © 2006 John Wiley & Sons, Inc.

many rivers in the United States are currently above 4 mg/l phosphate and are thus ripe candidates for eutrophication.[2] Algal blooms generally do not occur at phosphate concentrations less than 0.018 mg/l. Much of the phosphate in lakes and rivers is in the form of fine particulate material, which is either washed from the soil during rainfall or discharged from various point sources including wastewater treatment plants and feedlots.

Phosphate removed in the wastewater treatment plant is often re-released when the biosolids (sludge) is land applied for disposal. Because of the adoption of the ban on phosphates in laundry detergents, other chemicals have been substituted.[3] The quantity of phosphate has been reduced, but the European Chemical Council makes a persuasive case for partially lifting the ban on phosphates because they are better for the environment and biosolids management when all factors are considered.[4] This includes sludge volume and chemicals remaining in the sludge.

The largest current source of phosphate in the environment currently appears to be from agricultural applications, including Concentrated Animal Feeding Operations (CAFOs). The regulatory response to agricultural and feedlot applications of phosphate appears to be focusing on the control of phosphorus in animal feed to concentrations utilized by the animals, and agronomic application of phosphate in crops, limiting the phosphate application to the amount required by the specific crops.

In many communities the water quality limits dictate the maximum daily load of phosphate discharged to the stream either by concentration or by total load. The current trend is to regulate on a watershed basis and develop a TMDL (Total Maximum Daily Load) for that watershed, considering all sources within the watershed. In many communities, which are now

[2]Litke, David W. Review of Phosphorus Control measures in the United States and their effect on Water Quality. USGS Water-Resources Investigation Report 99-4007. Available in pdf format at:http://water.usgs.gov/nawqa/nutrients/pubs/wri99-4007.

[3]Typical laundry formulations for nonphosphate detergents include up to 25% zeolite (sodium aluminium silicate), 15%–20% sodium carbonate, 15% surfactants, 13%–15% of surfactants, and 13%–18% of sodium perborate, 2%–5% sodium sulphate, and about 1% enzymes. Source: CEEP.

[4]CEEP is a working group within the Central European Chemical Industry Council, located in Bruxellex, Belgium. Their publications have analyzed the character and performance of phosphate detergents and found that phosphates are not as harmful to the environment as was first thought, and that they are a very effective detergent builder, outperforming their substitutes. The other findings include the fact that phosphate detergents account for about 25% of the total P in the influent of the wastewater treatment plant (the balance is from human sources), and the overall volume of wastewater sludge generated when phosphates are used is significantly lower than when other forms of detergent substitutes are utilized.

instituting phosphate limits it is not uncommon to find discharge water quality limits of the order of 0.2 mg/l total phosphate. This is often a difficult goal to attain without advanced treatment.

The principal form of phosphate is polyphosphate, a compound that is incorporated into the ADP in cells and as phospholipids in cellular materials. Domestic wastewaters contain between 6 and 20 mg/l total phosphate, of which only 10–15% is organic phosphate. The balance is inorganic phosphate that is generally obtained from detergents and other fertilizer sources.

To understand the effects of phosphate contamination, look at the two sets of illustration given in Figure 11.1. These were taken from the following site, which is a good summary for phosphate issues: http://www.nhm.ac.uk/mineralogy/phos/index.htm.

The photos demonstrate clearly the need to remove phosphate.

There are two principal methods of removing phosphates, biological and chemical. The chemical method is through precipitation with metallic ions. The biological method is discussed in the following section, followed by the chemical method.

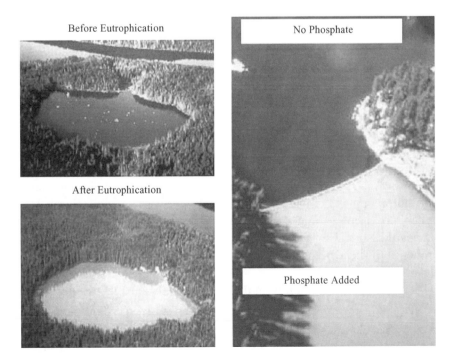

FIGURE 11.1 Several lakes showing the effects of algal blooms because of phosphorous levels in the lake.

BIOLOGICAL PHOSPHOROUS REMOVAL

Revisiting the C:N:P ratio discussion from earlier tells us that it takes about 100 units of carbon to remove one unit phosphorus. The "secret" to natural phosphorous removal appears to be the same method used in total ammonia and nitrogen control—anoxic treatment. The principal organism for this removal is the bacterium *acinetobacter*, a ubiquitous bacterium. There is uncertainty whether the bacterium is stressed by an anoxic period, or whether the anoxic period allows it to utilize other carbon sources and prepare for growth and then store up excess phosphate for future growth. The consensus is that the organisms use acetate and short-chain fatty acids to store polyphosphates as poly-β-hydroxybutyrate (an acid polymer). The exact mechanism is tied up with the production and regeneration of ADP within the cellular material, and it involves the ADP/ATP chain within the bacteria, and it is an extremely complex subject, which will add little to this discussion.[5] This uptake phenomenon was also known as "luxuriant uptake of phosphorus" in some of the earlier literature because the bacteria involved actually acquire more phosphorus than they require for growth. This is done principally through extracellular enzymes, and the bacteria stores the phosphorus until it is required for growth or respiration.

In a wastewater treatment plant, phosphorous removal and nitrogen removal do not occur simultaneously. Nitrate appears to interfere with phosphorous uptake, and phosphorus will not be removed until the nitrate is gone. Phosphate removal requires true anaerobic conditions, which occur only when there is no other oxygen donor.

If nitrate reduction and phosphorous uptake are to take place then an additional source of carbon is required. As was discussed previously, the carbon source can be either an added sugar, an alcohol, or the carbon in the entering wastestream. In the anaerobic process, the bacteria first release their extracellular phosphorus and then uptake more than they released.

As shown in Figure 11.2, there are a number of configurations possible for phosphorous uptake streams. There are even several modifications of the BardenPho® process that will allow a variety of options and adaptations. Perhaps the simplest one is the first one shown in Figure 11.2, where the aeration at the head of the main tank is turned off and the system is allowed to go partially anaerobic.

There are several variations on this process with some interesting caveats the operation. Recycled sludge from an aerobic tank contains dissolved

[5]See Orhon D, Artna N. Modeling of Activated Sludge Systems. Technomic Press, 1994.

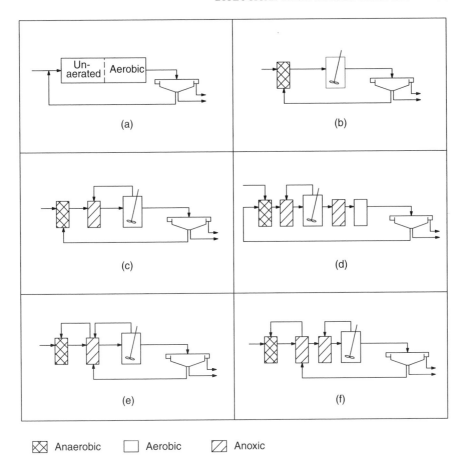

Anaerobic   Aerobic   Anoxic

Principal Phosphorous Removal Systems: (a) modified activated sludge system, (b) "Phoredux" two stage system, (c) "Phoredux" three stage system, (d) Bardenpho System (modified by "Phoredux"), (e) System developed by University of CapeTown (South Africa), and (f) UCT process modified for biological P and N removal.

FIGURE 11.2   System of phosphorous removals. Note that the similarities between these processes and the ones in Figure 10.5.

oxygen generally between 2 mg/l and a high of about 6 mg/l, although the latter figure represents poor practice because it is wasteful. It will require some time for the bacteria in the entering waste to consume the oxygen in the return sludge and recycled aeration return. Thus the initial mixing will not be anaerobic for some time. That is highly dependent upon the amount of dissolved oxygen in the recycle lines as well as the biosolids in those lines, and achieving anaerobic or fermentation conditions may require between 2 and 5 h. If the wastewater treatment plant is designed with long narrow tanks, it is comparable to plug flow and it will be possible to attain anaerobic conditions if the aeration can be adequately controlled. By comparison, use of circular tanks for aeration will require separate tankage.

Work by Jiang et al. (2005) indicates that there is an optimum detention size for the anaerobic tankage to minimize total plant cost.[6] That size is approximately 3 h, and is independent of the size of the aeration tankage with regard to the performance of the phosphorous removal of the system.

BardenPho® Phostrip® Processes

While working on bench-scale nitrogen removal systems in South Africa in the early 1970s, Barnard observed phosphorous removal when the initial zone of the bioreactor was anaerobic (no dissolved oxygen and no nitrate oxygen present). This finding convinced Barnard to work further on the application and develop several biological nitrogen and phosphorous removal systems in South Africa. The largest of these systems are in Pretoria, Cape Town and Johannesburg. These systems were designed to operate at relatively long hydraulic retention times (~18 to ~24 h) and high sludge retention times (SRT or sludge age) (>16 days).

The process was introduced to the United States in the early 1980s, at Palmetto. Cold weather adaptations of the process were installed at a 23,000 m^3/day plant in Kelowna, British Columbia. It should be noted that cold weather application is a significant accomplishment because of the severe effects of cold weather in decreasing the rate of nitrification and denitrification. Since then there has been an incorporation of a number of modifications in the technology to make the process more cost-effective and adaptable to a wide variety of applications and climates.

The BardenPho® and similar processes start with high SRT-extended aeration systems, which generate a highly nitrified effluent. The effluent is then stirred in an anoxic tank followed by anaerobic stirred tankage. The process removes both nitrogen and phosphorus. Design is somewhat complex, and there are a number of variations of the process.

A schematic of the BardenPho process is shown in Figure 11.3. Note that the influent is from a secondary treatment process:

[6]Jiang F, Beck MB, Cummungs RG, Rowles K, Russell D. Estimation of Costs of Phosphorous Removal in Wastewater Treatment Facilities: Adaptation of Existing Facilities. Water Policy Working Paper #2005-011 February, 2005. Available at http://h2opolicycenter.org/pdf_documents/W2005011.pdf.

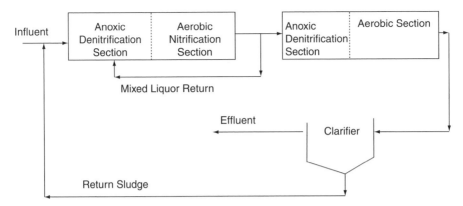

FIGURE 11.3   Bardenpho Process using two tanks. Note that tanks need to be long and narrow approaching plug flow. Compare this drawing to Figure 10.7.

The modified Ludzack–Ettinger process also relies on a secondary treatment system before it is applied. As an adapted system, the piping is generally re-routed so that the existing clarifier can be used for the final solids removal (Fig. 11.4).

Another process for removal of phosphorus is the Phostrip process:

The Phostrip system receives the effluent from the activated sludge system. In it a small portion of the settled sludge from the activated sludge clarifiers is diverted to an anoxic phosphorous strip tank where it will be held for several hours in an anaerobic condition and enhanced with acetic acid to induce the sludge to release its phosphorus.

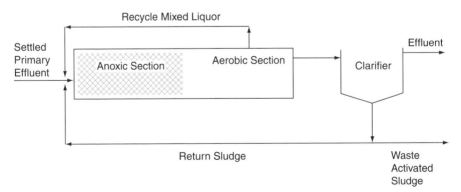

FIGURE 11.4   Modified Ludzack–Ettinger Process for Phosphate Removal. Requires long narrow tanks for maintenance of plug flow, and in the Anoxic Section the Waste Activated Sludge is returned from the clarifier as well as recycle mixed liquor from the aerobic section. The latter needs to be closely controlled if anerobic conditions are to be maintained.

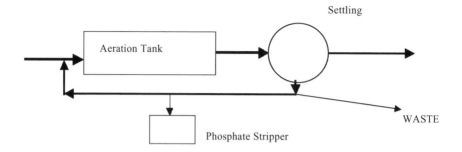

PHOSTRIP PROCESS

FIGURE 11.5   Schematic of the Phostrip process. The sludge is treated chemically to remove phosphorous before it is returned to the aeration tank.

The low phosphorous sludge is then washed with a portion of the clarifier effluent. The elutriate from the washing process contains 60 mg/l–80 mg/l of phosphorus, and it is reacted with lime to remove the phosphorus. The stripper overflow after treatment is very low in phosphorus, and it, along with the phosphorous poor sludge is returned to the aeration tank.

This is shown in Figure 11.5:

CHEMICAL PHOSPHOROUS REMOVAL

Chemical phosphorous removal is precipitation with a divalent cation. The phosphates are insoluble in varying degrees depending upon the cation used. The most common compounds used for phosphorous precipitation include lime, limestone, magnesium, iron salts, aluminum, and alum salts. Of these, alum, ferric chloride, and lime are the most common. Ferric chloride is the most popular because it is the most effective for the price. Alum creates a lighter floc, which settles more slowly, and lime has the disadvantage of raising the pH and possibly interfering with other process applications, generating a higher volume of sludge, and potentially creating pH control problems with the effluent (Fig. 11.6).

Common dosages are stoichiometric, and the advantage of chemical additions is that aside from tankage, pumps, and mixing required for the chemical application (very small capital cost when compared with the initial investment for the treatment plant cost or upgrades for enhancements to the biological processes), the removal of phosphorus can be stepped to achieve almost any desired effluent level.

According to various studies and estimates, domestic sewage contains between 2 and 14 mg/l of total phosphorus, averaging about 6.34 mg/l—of

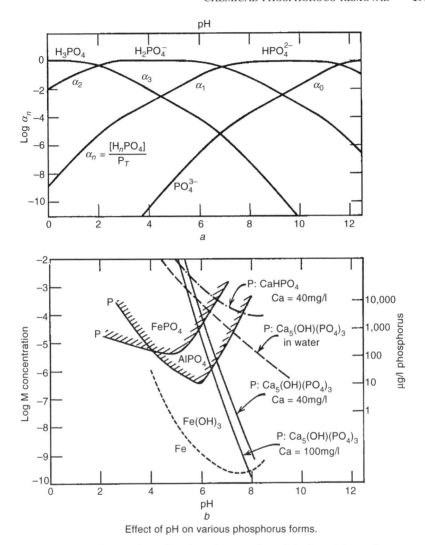

Effect of pH on various phosphorus forms.

FIGURE 11.6 Various phosphate forms and precipitation with metals.

that about 3 mg/l is soluble ortho-phosphorus.[7] A normal wastewater treatment plant will remove about 61% of the total phosphorous load and will provide an effluent of about 2–2.5 mg/l under normal circumstances.

[7]EU Cost 624 Study. Available at http://www.ensic.u-nancy.fr/COSTWWTP and Jiang F, Beck MB, Cummungs RG, Rowles K, Russell D. Estimation of Costs of Phosphorous Removal in Wastewater Treatment Facilities: Adaptation of Existing Facilities. Water Policy Working Paper #2005-011 February, 2005. Available at http://h2opolicycenter.org/pdf_documents/W2005011.pdf.

TABLE 11.1   Precipitation Reaction for Various Phosphate Forms (Solubility of Phosphates and Condensed Phosphates)

Reactions	Log_{10} Equilibrium Constant
$Ca_5OH(PO_4)_3 = 5Ca^{2+} + 3PO_4^{3-} + OH^-$	-55.6
$Ca_5OH(PO_4)_3 + 3H_2O = 2(Ca_2HPO_4(OH)_2)$ $+Ca^{2+} + HPO_4^{2-}$	-8.5
$Ca_2HPO_4(OH)_2 = 2Ca^{2+} + HPO_4^{2-} + 2OH^{2-}$	-27
$CaHPO_4 = Ca^{2+} + HPO_4^{2-}$	-7
$FePO_4 = Fe^{3+} + PO_4^{3-}$	-23
$AlPO_4 = Al^{3+} + PO_4^{3-}$	-21
$Ca_2P_2O_7 = Ca^{2+} + CaP_2O_7^{2-}$	-7.9

The cost of implementing chemical precipitation is approximately one-half of the cost of implementing phosphorous removal by biological means. However, as the phosphorous limit decreases, the cost differential between various methods decreases because of the need to add additional process equipment (filtration) to remove fine particulate solids containing phosphorus. It is also not uncommon for a wastewater treatment plant to be adding chemical (ferric chloride, alum, polymer, or any combination of these) worth of several hundred dollars per day to the plant effluent for removal of phosphorus.[8]

One of the definitive works on the subject of *Process Design Manual for Phosphorous Removal* was prepared by Black and Veach Engineers for the USEPA, in 1971. This publication is still available from the NTIS. The publication is a bit one dated but still contains useful data for design parameters.

The following precipitation reactions are important (Table 11.1).[9]

Most if not all the data above are generated in a laboratory in pure water under laboratory conditions. The conditions in a clarifier full of biologically treated wastewater are substantially different from those encountered in the laboratory. The chemical addition for phosphate removal appears to be independent of the location where the precipitants are added. In some cases, it may be just as easy to remove phosphate in the primary clarifier where the increase in sludge volume may not be as significant as it may be in the final clarifier.

---

[8]A recent study (2005) indicated that municipal wastewater treatment plants on the Etowah River (North of Atlanta, GA) were adding between 110 and 160 gallons/day of commercial grade ferric chloride ($FeCl_3$) or about 4–5 mg/l to obtain an effluent quality of approximately 0.3 mg/l total P. Internal notes and study for University of Georgia—Warnell School of Forestry.
[9]Stumm, Morgan. Aquatic Chemistry. John Wiley & Sons, 1996, ISBN 0471-81185-4.

The principal concern is the cost of the materials. Waste pickling acid has been used as a source of ferric chloride, as have been a number of other products. It is possible to introduce directly the metal ions into the wastewater by using direct current and by electrolytically generating the ions using sacrificial cathodes.

A final note: Handle the chemicals carefully. Each has its own strong and weak points with respect to its use and your treatment options. Consider chemical staining, chemical handling, chemical toxicity, and other things that you would normally look at when you handle chemicals. If lime or hydrated lime is used for precipitation, you may also want to recarbonate the wastewater to add back alkalinity and lower the pH to normal treatment ranges.

A good source for chemical information is http://www.siri.org/msds/. Or go to www.google.com or any of the other search engines and search for Material Data Safety Sheets (MSDS). The database is huge but remember that MSDS are designed to provide only the basic information and not to answer some of the questions you may want to know about the chemical properties.

12

ANAEROBIC TREATMENT

Basic anaerobic processes
Anaerobic pretreatment
Sludge digestion
Sludge treatment
Anaerobic digester model ADM1

BASIC ANAEROBIC PROCESSES

Anaerobic processes are those that occur, by definition, without oxygen. It is different from an anoxic process because it is a reducing environment as opposed to an environment without oxygen. Both processes are anoxic, but anaerobic is a reducing environment beyond anoxic where oxidation reduction potential (ORP) values are strongly negative and nitrate is reduced to ammonia and nitrogen gas, and sulfate (SO_3^{2-}) is reduced to hydrogen sulfide. ORP for some of the nitrogen reductions to proceed is around $-300\,mV$ to $-750\,mV$. Phosphate is also reduced, but because it is often transformed through the ADP–ATP chain, it is difficult to estimate the ORP for the process.

Anaerobic decomposition produces both organic acids and gas. It is a slower process but it develops about one-fourth of the biomass of an aerobic process, the principal ones being the production of organic acids and gas. Anaerobic treatment takes place relatively slowly, is often temperature- and highly toxin-sensitive, can be easily upset, and requires substantial mixing.

Practical Wastewater Treatment, by David L. Russell
Copyright © 2006 John Wiley & Sons, Inc.

Anaerobic treatment is used in two areas. The first and most common is the degradation of biosolids from the treatment process. The second is the reduction and treatment of high-strength wastewaters containing soluble degradable organic materials. High-strength organic wastes where the influent BOD or COD is well above 500 mg/l are often very good candidates for anaerobic treatment. This could include animal feedlot wastes, sugar processing wastes, petroleum wastes (if the toxicity is controlled), and many canning and dyestuff wastes where water-soluble organics are used in the process.

Most anaerobic treatment (solids digestion) takes place in two specific temperature ranges–mesophilic and thermophilic. The temperature ranges are of the order of 85°F–100°F (30°C–38°C) and 120°F–135°F (38°C–50°C), respectively. From personal experience, the total gas production is about the same either way, but it is generated much faster at the thermophilic range and falls off more quickly for a given batch feed. Organisms can be brought from one temperature range to the other if the temperature conversion is performed slowly enough. It is a matter of re-growing or re-acclimatizing the existing organisms. The process often takes a week or two of gradual temperature changes. Above the thermophilic range, the temperature effects often cause partial sterilization and loss of organic growth.

In anaerobic processes there are three parts

Fermentation of the wastes—conversion to acetates;

Acetogenesis—conversion to acids, formaldehyde, and hydrogen; and

Methanogenesis—conversion of formaldehyde, acetates, and acids to CO_2 and methane.

One of the principal challenges to anaerobic treatment is balancing the rates of growth. The acid-forming bacteria operate at about three times the rate of the methane-forming bacteria, and without a balanced microbial population the wastes will turn acidic and all methane production will stop. When anaerobic digesters are in "start up" mode, this condition can occur, and it is known as a "stuck digester." It is cured by the slow addition of alkaline buffers (dilute lime) to the mix. Strong alkalis can take the mix well out of the sludge range, where all activity stops.

Anaerobic fermentation can occur in the pH range of between 5.0 and about 9.0, while the bacterial methane operates in a much narrower range of between 6.5 and about 7.6, with the optimum range of about 7.0.

General formulations for anaerobic decomposition have been provided by Buswell for carbohydrates:[1]

$$C_nH_aO_b + (n - a/4 - b/2)H_2O \rightarrow (n/2 - a/8 + b/4)CO_2$$
$$+ (n/2 + a/8 - b/4)CH_4$$

Most of the bacterial acids formed are generally proprionic and acetic acids. Another researcher, McCardy and Murdoch (1963) estimate the following:

Amino and fatty acids	$A = 0.054\,F - 0.038\,M$
Carbohydrates	$A = 0.46\,F - 0.088\,M$
Nutrient broth	$A = 0.076\,F - 0.014\,M$

where A = biological solids accumulated; M = ML VSS; and F = COD utilized.[2]

Another way of looking at the same reactions is provided by McCardy and Rittman (2001):

$$C_nH_aO_bN_c + (2n + c - b - 0.45df_s - 0.25f_e)H_2O \rightarrow \rightarrow$$
$$0.125df_eCH_4 + (n - c - 0.2df_s - 0.125df_e)CO_2 + 0.125df_eCH_4$$
$$+ (n - c - 0.20df_s - 0.125df_e)CO_2 + 0.05df_sC_5H_7O_2N$$
$$+ (c - 0.05df_s)NH^{4+} + (c - 0.05df_s)HCO^{3-}$$

where $d = 4n + a - 2b - 3c$ and f_s is the fraction of organic matter (COD or BOD) converted to cells and f_e is the portion converted to cellular energy, such that $f_e + f_s = 1$. Where f_s may be estimated from cell yield, varying for different compounds.[3] Typical values of f_s vary from 0.042 for fatty acids to 0.11 for methanol and 0.2 for carbohydrates. The value for proteins is 0.056.

The wastes must have a balanced feed, including freedom from high concentrations of salts, and relatively high levels of alkalinity must also be present to counteract the CO_2 generated. The following chart (Figure 12.1) is taken from a sludge digestion manual but it illustrates the point.

[1]Buswell AM, Mueller HF. Ind. Eng. Chem. 1952, Vol. 44, pp. 550–552 and Industrial Fermentations, New York: Chemical Publishing Company, 1954.
[2]McCardy PL, Murdoch W. JWPCF 1963, Vol. 35, pp. 1501–1516.
[3]McCardy, Perry and Rittman, Bruce. Environmental Technology, Principles and Application. New York: McGraw Hill, 2001, ISBN: 0072345535.

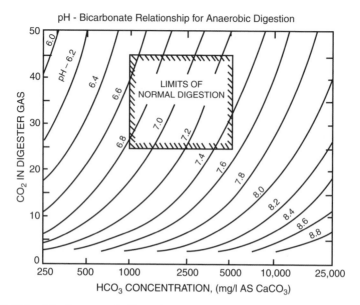

FIGURE 12.1   Bicarbonate and pH requirements for sludge digestion. *Source*: Process Design Manual for Sludge Treatment and Disposal, USEPA 1974.

ANAEROBIC PRETREATMENT

This is performed on a number of high-strength biodegradable waste materials. The following table taken from the study of McCardy and Eckenfelder as published in *Public Works Journal*, September–December 1964 Public Works Journal, V. 85, No. 3 is shown in Table 12.1. Note that the rates of loading are very high for the specific wastes, per unit volume of digester volume. The loading rate of 1 lb of X per 1000 of digester is equivalent to $0.0160 \, \text{kg/m}^3$.

In many instances, the completely mixed anaerobic reactor depends upon good mechanical agitation. A good reactor should have the following general elements:

1. provision for mixing;
2. allowance for gas handling;
3. temperature controls;
4. sampling ports;
5. solids removal system;
6. mechanical scum and hair breaking (note war story!);
7. climatic seals to insure anaerobic conditions maintained;
8. provisions for odor control of gasses and effluent;

TABLE 12.1 Anaerobic Pretreatment Design Parameters for Contact Stabilization and Actived Sludge Processes

Anaerobic Pretreatment of Industrial Wastes for Contact Stabilization and Activated Sludge Processes (Data from McCardy and Eckenfelder)

Anaerobic Pretreatment Before Activated Sludge Plant (Industrial Water Pollution Control, Wes Eckenfelder, Mc Graw Hill, 1966)

Waste	Hydraulic Retention Time in Days	Digestion Temperature °F	Raw Waste BOD$_5$ mg/l	Pounds of VSS Added per 1000 of Digester Volume	Pounds of BOD$_5$ Added per 1000 of Digester Volume	Percent Removal Basis is Expressed Same as Material Added
Pea Blanching	3.5	131		700		83
Pea Blanching	6.0	99		400		85
Winery		97		200		87
Rye Fermentation	2	130		930 (Total Solids)		54
Corn Fermentation	4	130		330 (Total Solids)		76
Whey Wastes	29	130		150 (Total Solids)		72
Acetic Acids	30	95		1370	975	90
Maise Starch	3.3	73	6280		110	88
Whiskey Distillery	6.2	92	25,000		250	95
Citrus	1.3	92	4600		214	87
Wine	2.0	92	23,400 (VSS)	730		85
Molasses	3.8	92	32,800 (VSS)	546		65
Meat Packing	0.5	75–92	1110–1380		131–156	91–95

This is true of sludge digestion as well as anaerobic filtration.

Some anaerobic filters are established as attached growth or sludge blanket type units with either fluidized beds or fixed media to give a greater density of organisms. This is often met with mixed results, including various types of plugging and blinding from too much organic growth and from other materials in the waste. However, the greater surface area of a packed bed anaerobic filter often offsets the operational difficulties and allows for greater density of organisms in a smaller reactor volume, thus saving capital costs.

The use of activated carbon as an attachment medium for the fluidized bed reactor has been a more recent trend in anaerobic reactors. The activated carbon will adsorb a number of toxic compounds such as phenols and allow the bacteria time to acclimatize them to the lower relative density of the activated carbon.

One author has suggested separation of the two types of growth into an acid reactor and a methane reactor. The reactors would be of different sizes. This approach has also met with some mixed results and has not been universally adapted. This approach has been used primarily with sludge digestion.

SLUDGE DIGESTION

Most sludge digestion occurs after an aerobic biological treatment. The excess solids are loaded into an anaerobic reactor for digestion. The holding times are generally a minimum of 30 days and some, depending upon temperature and gas production, are as high as 180 days.

Frequently the sludge is prethickened before it is sent to the digester. On larger systems this will increase the amount of solid in the feed from the range of 1–3% to 7–10% or more if achievable. Thickening can be accomplished in a variety of ways, including gravity settling with gentle stirring (conventional thickening) and flotation. The point is to increase the solids feed and decrease the total amount of liquid stored. The upper limit is the ability to pump and stir the solids.

The digester loading rates are generally either conventional rate or high rate. The loading rates are low rate 40–100 lb/1000 lb/ft^3/day and high rate 150–400 lb/ft^3/day, with average solids retention times of 30–60 days for low rate and 10 days–20 days for high rate. Two types of digesters are shown in Figures 12.2 and 12.3.[4]

---

[4]Reid-Crowther Web site—no longer posted.

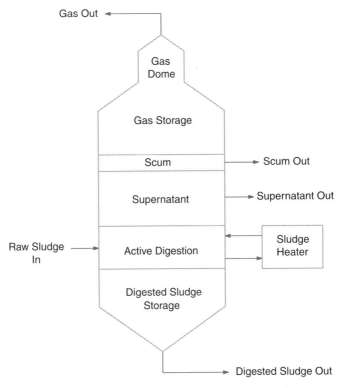

FIGURE 12.2 Single stage conventional anaerobic digester.

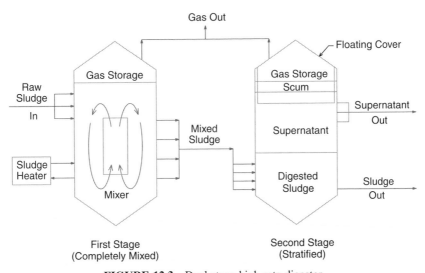

FIGURE 12.3 Dual stage high rate digester.

SLUDGE TREATMENT

After digestion, the sludge is mostly inactive but contains substantial quantities of ammonia. It is often further processed by air drying on sand beds, in greenhouses, and by filter pressing and with thermal and other types of treatment including composting. Sludge drying beds are noted for growing tomatoes indoor as the seeds are unaffected by the digestion process.

Recently the trend has been to dispose of sludges by direct soil incorporation in landfarming, composting, and with polymer treatment and filter press, in landfills. This landfilling is wasteful and expensive and does little to help the landfill unless the landfill is to be converted into a large anaerobic or aerobic reactor. In fact, there is substantial evidence that most landfills are extremely poorly stirred anaerobic digesters. Enhancements to landfills where cells are opened up to recirculation of the leachate and even conversion of landfills to aerobic biological reactors have proved successful in reducing the volume in the landfill and in stabilizing the landfill.

ANAEROBIC DIGESTER MODEL ADM1

In 2004, the IWA published an comprehensive mathematical model of the anaerobic digestion process, ADM1. Almost as soon as the model was released, there were reports about mistakes in the carbon balance, which many of the modeling firms and the IWA have subsequently corrected. The corrected versions are working satisfactorily, and because of the model, there is a current EU COST 624 Working Group on "Optimal Management of Wastewater Systems" Working Group N°1 "Plant Operation" which has been working since at least 2001 to provide plant wide models which now include the operation of an anaerobic digester.

The model has been upgraded and extended by various researchers to include sulfate-reducing processes and nitrate reduction processes, and other multiple reaction stoichiometry, microbial growth kinetics, conventional material balances for ideally mixed reactor, liquid-gas interactions, and liquid-phase equilibrium chemistry. As it is written, the model has at least 26 dynamic state variables, modeling 19 biokinetic chemical processes, and incorporates 3 gas-liquid kinetic transfer processes.

Reports from researchers using ADM1 have indicated good correlation between the model and performance in laboratory and other applications. As of 2006, the EU even has a specification for a "Plant Wide Model" which includes a dataset and can be used to help optimize all phases of plant operations, including the anaerobic digester. The ADM1 model is also being used to research anaerobic pretreatment for wastewater processes.

13

MICRO/ULTRAFILTRATION

Introduction to membrane separations
  and microfiltration
Design values
Process selection

INTRODUCTION TO MEMBRANE SEPARATIONS
AND MICROFILTRATION

Microfiltration and ultrafiltration are forms of superfine screening. The universe of microfiltration varies from microfiltration to nanofiltration and reverse osmosis. The commonality between these types lies in the uses of permeable membranes. The membranes can be made from ceramics, foils, etched polymers, and natural and synthetic compounds. The common materials of construction include titanium or zirconium dioxides (ceramics), and cellulose acetate, polyamide, polypropylene, polysulfone, polytetrafluoroethylene, polyvinylidene fluoride. Each has its own properties and specific resistances to heat, bacterial attack, corrosion, and abrasion. Consult the manufacturer of specific membranes for information on the properties and resistances of the membrane materials.

It is often helpful to think of membranes in terms of screens rather than filters because the membrane does not deliberately build up a cake in the same way that a filter does, and once the pores are plugged, the head losses across the membrane climb steeply.

Practical Wastewater Treatment, by David L. Russell
Copyright © 2006 John Wiley & Sons, Inc.

TABLE 13.1   Membrane Separation Properties and Performance

Type of Membrane	Separation Mechanism	Pore Size (microns)	Molecular weight (amu or Da)	Operating Pressures (psi)
Reverse osmosis	Screening and diffusion	<0.001	100–200	600–1500
Nanofiltration	Screening and diffusion	0.001–0.01	300–1000	50–250
Ultrafiltration	Screening	0.01–0.1	1000–100000	3–80
Microfiltration	Screening	0.1–20	Over 100000	1–30 (or vacuum)

The sizes of membrane pore vary greatly, as do the pressure drops. Starting with the finest sizes and working up the following is generally applicable to all membranes:

One dalton is one-twelfth the weight of a carbon atom, as defined by convention in 1960 and is approximately equal to $1.66053873 \times 10^{-24}$ g.

Membranes are not always homogeneous or isentropic. Some membranes are designed to have a different density and pore gradient from one side to the other, and others are composed of layers of different materials for a very fine cleaning. Membranes often require careful cleaning or prefiltration to prevent their clogging. The finer the membrane, and the higher the pressure, the more likely that the membrane will require prefiltration and conditioning if it is to enjoy long life.

Membranes are not something over which the design engineer has much control. It is often more a matter of looking at the equipment available and sifting through the manufacturer's claims regarding their equipment and then selecting the best guarantee and balancing that against the price of the equipment and anticipated performance. In one very large plant in Gwinnett County, Georgia, a 60 million gallon/day advanced wastewater treatment plant (227125 M^3/day), the County and the design engineer arranged a side-by-side comparison of membranes from different manufacturers to select the best equipment. This is a very good idea until the technology becomes more widely used and proven.

The key to establishing long membrane life is crossflow cleaning of the membrane coupled with frequent backflushing or chemical cleaning, depending upon the type of membrane. If the contaminated liquid flows across the membrane and not normal to it, the membrane filtration run is improved because the *crossflow* removes the solid buildup, which would plug the pores. For certain types of microfiltration membranes, such as those becoming more frequently used in wastewater treatment plants, the tubular hollow membrane is directly submerged in the aeration portion of the plant

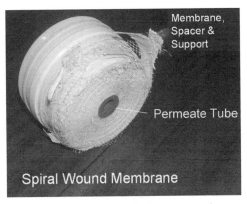

FIGURE 13.1   Partially disassembled spiral wound membrane.[1] Note: Mesh and backing to membrane.

where it is in contact with bacteria, viri, and protozoans in the wastewater as well as colloidal solids and suspended materials. For the "naked" tubular membrane applications, the manufacturers have instituted frequent back-flushing and an "air bump", which not only shakes and scours the membrane but also promotes knocking off of the fouling layers. This coupled with backflushing and pulsing of the membranes helps keep the pores open.

In wastewater applications, depending upon the process selected, the cleaning compounds may include sodium hypochlorite, hydrochloric acid, ozone, oxalic acid, hydrogen peroxide, and citric acid. The principal use of the acids is to clean out the carbonate buildup in the membrane pores. Cleaning cycles vary from a few hours to 24 h, depending upon the severity of the plugging. The backflushing and back-pulsing can take from a few minutes to an hour once per day. The chemical cleaning of the membranes is performed weekly to monthly, depending upon the need and the manufacturer's instructions.

A type of membrane design is shown in Figures 13.1 and 13.2.

The older style membrane tubes have a diameter between 2 and 5 cm. In the newer spaghetti strand designs, the tubes are of the order of 0.5–1 mm diameter with the same hollow core. In larger configurations, the feed is sometimes from the inside of the tube into a shell. The smaller diameter tubes use an outside feed and operate on pressure differential. In wastewater treatment plant configurations, the membrane tube is often in contact with the water and a vacuum is pulled over the tube. In either case the membrane is sealed into the base with an epoxy seal and a machined slot for an O-Ring gasket.

Most of the time pretreatment of the water is necessary. Even in wastewater plants, prescreening to remove hair, and free fats and oils, is

[1]www.osmonics.com, p. 715.

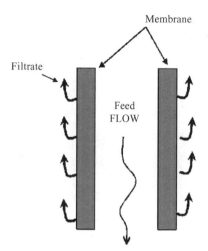

FIGURE 13.2   Feed/flow path for a tubular membrane. The flow pattern for the spaghetti strand membranes is outside to inside.

necessary. Depending upon the application, some posttreatment may also be necessary.

Table 13.2 presents typical effluent quality from a microfiltration system.[2]

There is still a large caveat. The effluent water is still not sterile because it will contain viri, and protozoan cysts, and other things with molecular weights less than 100,000 Da. It must also be noted that, depending upon the size of the filter used, the suspended solids may be strained out, at sizes less than 0.45 microns.[3]

TABLE 13.2   Typical Effluent Concentration after Membrane Filteration

Parameter	Units	Approxmiate Effluent Quality
BOD	mg/l	<2–5
Total organic carbon	mg/l as C	5–25
Total kjelldahl nitrogen	mg/l as N	5–30
Total phosphorous	mg/l as P	0.1–1.8
Iron	mg/l	0–0.2
Total suspended solids	mg/l	BDL
Fecal coliform	No./100 ml	2–3

[2]Water Environment Federation. Membrane Systems for Wastewater Treatment Table 4.2, New York: McGraw Hill, 2005, ISBN 0-07-146419-0.
[3]Many of the current microfiltration systems are being installed with pore sizes of 0.4 microns. The definition of suspended solids uses a 0.45 micron filter so the TSS is zero. It is also important to remember that when modeling a process using the ASM models, the definition of soluble materials is anything that passes 0.2 micron.

Membrane separations are usually less costly in capital and operation costs than other forms of separation, and a paper on costs of wastewater treatment brings out this point.[4]

DESIGN VALUES

The design values for membrane microfilters are of the order of 30–45 $L/H/M^2$ or 16–25 gallons/day/ft^2 for straight microfiltration on immersed activated sludge systems, and between 50 and 60 $L/H/M^2$ for relatively clean wastewater after treatment or pretreatment by a clarifier. A word of caution is in order. Many wastewater systems have a diurnal variation factor of 2.5:1 or more depending upon the amount of inflow and infiltration and other factors. The design for the system should be for the peak flow because the systems do not take surges well. Some manufacturers recommend that backwashing of the membrane filter systems should not be scheduled for the time of day when the flow is the highest.

PROCESS SELECTION

The following summary can help with the types of separation:

Ultrafilteration Membrane Selection

Micro/Ultrafilteration membranes should have the following:

High fluxes
Sharp molecular weight (MW) cutoff
Good mechanical, chemical, and thermal stability
High-life expectancy

The most widely used polymers are cellulose acetate (CA), aromatic polyamides, polysulfones, and polyacrylonitrile-poly(vinyl chloride) co-polymers.

[4]Jiang F, Beck MB, Cummungs RG, Rowles K, Russell D. Estimation of Costs of Phosphorous Removal in Wastewater Treatment Facilities: Adaptation of Existing Facilities. Water Policy Working Paper #2005-011 February, 2005. Available at http://h2opolicycenter.-org/pdf_ documents/W2005011.pdf and Jiang F, Beck MB, Cummungs RG, Rowles K, Russell D. Estimation of Costs of Phosphorous Removal in Wastewater Treatment Facilities, DeNovo. Water Policy Working Paper #2004-10.

Cellulose Acetate Membranes

Advantages
  Can be produced in a wide range of pore sizes
  Obtain relatively high fluxes
Disadvantages
  Mechanically weak and thermally unstable
  pH operating range of 4–8
  full temperature of 35°C
  Susceptible to bacterial attack

Polysulfone Membranes

Advantages
  Excellent chemical stability
  pH range of 0–14
Disadvantages
  Certain materials are adsorbed at the membrane surface
  Hard to achieve low MW cutoff characteristics

Polyamide Membranes

Advantages
  Low MW cutoff and good flux can be achieved
  Excellent mechanical strength and thermal stability
Disadvantages
  Sensitive to chlorine attack at low concentrations
  Adsorb certain materials at surface

Polyacrylonitrile Membranes

Advantages
  Can be dried completely and re-wetted without changing filtration
    characteristics
Disadvantages
  MW cutoffs above 30,000
  Low mechanical strength

Ultrafiltration Modules

Choice of module design is as important as the membrane material. At
sufficiently high pressures a thin gel layer forms, caused by concentration

polarization, at the membrane surface. This gel layer can alter properties of the ultrafilteration membrane. Flow control at the membrane surface must be taken care of.

Tubular Module   Pressurized feed flows into the center; the permeate flows through the membranes and is collected in the outer shell. These tubes can be installed in parallel or in series.

Advantages

Tolerant toward suspended solids
Easily mechanically cleaned
Control of concentration polarization effects
Adjustments overfeed flow velocity over a wide range

Disadvantages

High investment costs and operating costs
Low membrane surface area to system volume

Plate and Frame Module   Some membranes are configured in a stacked array like a filter press, with the same spacers and membrane supports shown in the spiral wound membrane shown above. The feed is from inside out and is channeled across the membrane.

Advantages

Large membrane surface are a per unit volume
Generally low operating costs

Disadvantages

Control of concentration polarization is more difficult
Plugging of feed flow can be problematic

Spiral Wound   There are limited uses of spiral wound modules in ultrafilteration; however, it is the most widely used type in reverse osmosis (RO).

Advantages

Membrane surface area per unit volume is high
Capital and operating costs are low

Disadvantages

Hard to control concentration polarization

Capillary The capillary module has a large number of membrane capillaries with diameters from 0.5 to 1.5 mm. The feed is passed through the center of each capillary and the filtrate permeates the walls.

Advantages

Low capital costs
Good feed flow control
Large membrane surface area per unit volume

Disadvantages

Hard to control concentration polarization

Rod Membrane Although they are similar to hollow fiber in RO modules, they have a grooved rod coated with asymmetric membrane. In this module the feed is fed to the outer side of the rods and the permeate is collected in the middle by grooves in the rod.

Advantages

High membrane surface area per unit volume

Disadvantages

Feed flow control is not good

Newer Applications As indicated in Chapter 6, the past few years have seen a relatively new use for membranes. The materials of construction are continually getting better, and now it is not difficult to find that membranes are being used as a replacement for a final clarifier. It may be worthwhile to look at the membranes in view of what we have learned about them. The membranes being used as wastewater treatment devices are relatively chemically inert. In fact, they are recommended for direct contact with bacteria. They have a pore diameter of about 0.45 microns, and a modest pressure loss of about 3–10 psi. The buildup on the membranes is prevented, according to the manufacturer, by a periodic air bump or backwash surge and the flexing of the membranes in the water. The process guarantees on the membranes are unconditional 3–5 years. The life of a wastewater treatment plant (economic life) is about 20 years, and the capital cost of the clarifier can be a substantial portion of the equipment.

Clarifiers are about half the cost of a membrane system and the cost factors increase according to the following equations, where Q is in million gallons per day:

$$\text{Clarifier capital cost} = 2.94 \times 10^{+5}Q^{+0.7}$$

$$\text{Membrane system capital cost} = 0.657 \times 10^{+6}Q^{+1.57}$$

$$\text{Sand filtration capital cost} = 7.718 \times 10^{+5}Q^{+0.74}$$

The source for this data is a study on wastewater treatment plant costs published in 2005 by the University of Georgia, Warnell School of Forest Resources and Georgia State University. Membrane systems have a higher initial cost and a higher capital cost than clarifiers; though initially they are slightly less costly than sand filters, as the flow increases the costs quickly mount.

The membranes also are being recommended for use in lieu of straight filtration even where chemical precipitation is being used. The backwash and

FIGURE 13.3   Cut away photo of spaghetti strand membranes system.

(air) bump rate on the membrane is increased slightly for those applications. If the membrane becomes plugged, it is removed and backwashed with either citric or acetic acids to restore its life. It also has the advantage that it can produce an effluent that has no effluent suspended solids.

The following photographs in Figures 13.3 and 13.4 will help us to characterize the membrane and the type of construction system.

FIGURE 13.4 Spaghetti strand membrane clarifier in operation (taken at WEFTEC 2003).

14

REVERSE OSMOSIS

Introduction
Mass transfer theory
Membrane selection
Membrane materials
Membrane configurations
RO design considerations
Design parameters

INTRODUCTION

Reverse Osmosis (RO) can remove more than 99% of all dissolved minerals and organic compounds, as well as biological and colloidal suspended matter, from water.[1] RO is useful in wastewater and process water treatment because it can be applied to each individual process and, therefore, to each individual separation problem. Additionally, recovery is often a plausible option because there is no chemical or thermal degradation.

MASS TRANSFER THEORY

RO is a process where the natural flow of fluid across a semipermeable membrane is reversed by applying pressure to the concentrated solution. When the applied pressure is greater than the natural osmotic pressure, the solvent will flow through the membrane to form a dilute solution on the

---

[1]American Waterworks Association Staff. Reverse Osmosis and Nanofiltration. Denver: American Waterworks Association.

Practical Wastewater Treatment, by David L. Russell
Copyright © 2006 John Wiley & Sons, Inc.

opposite side and a more concentrated solution on the side where the pressure is applied.

The temperature of the solution, membrane properties, and difference in applied and osmotic pressures all affect the flux of water across the membrane.

To calculate the flux of component A across the membrane, N_A, the following equation is used:[2]

$$N_A = P_A\left(\frac{\Delta\Phi}{L}\right)$$

where N_A = flux of A through the membrane (mass/time-length^2); P_A = permeability of A (mass-length/time-force); $\Delta\Phi$ = driving force of A across the membrane. This can be either a difference in concentration or a difference in pressure (mass/length^2 or force/length^2) and L = membrane thickness (length)

To obtain the osmotic pressure use the following equation:

$$\pi = C_s RT$$

where π = osmotic pressure (force/length^2); C_s = concentration of solutes in solution (moles/length^3); R = ideal gas constant (force-length/mass-temp); T = absolute temperature (°K or °R).

MEMBRANE SELECTION

The process of membrane selection is most often left to the manufacturer, and the consultant is asked to evaluate and confirm the manufacturer's selections. Because the equipment comes as a package or a system rather than as individual components, the designer will be asked to recommend the type of membrane and then the type of equipment package from the same manufacturer.

There are three membrane properties that are important for an economically successful application in the order of their importance: (1) membrane selectivity, (2) membrane chemical stability, and (3) membrane permeation or flux rate. Of these, the flux rate is the least important because the cost of additional membrane surface to make up for lower flux rates is a minor component of cost.

---

[2]EPA Capsule Report. Reverse Osmosis Process. Cincinnati: Center for Environmental Research Association, 1996.

MEMBRANE MATERIALS

The ideal membrane has the following characteristics: (1) high water flux rates, (2) high salt rejection, (3) high resistance or tolerance to chlorine and oxidants, (4) high resistance to biological attack, (5) high resistance to colloidal and suspended material fouling, (6) inexpensiveness, (7) easiness to form into films or hollow fibers, (8) high chemical and physical strength (resistant to high pressures), (9) high chemical stability (ability to retain chemical properties under a variety of conditions), and (10) high thermal stability (ability to withstand high temperatures without deforming or losing shape.)[3]

There are three main, different types of membrane materials available for RO: cellulose acetate (CA), aromatic polyamide (aramid), and thin film composites (TFCs).

Cellulose acetate is widely used, has low cost, and has the ability to withstand continuous exposure to low levels of chlorine. On the minus side, it tends to hydrolyze with time, has relatively poor chemical stability, and requires a pH range between 4.0 and 6.5, and is subject to biological attack. The upper limit of temperature is approximately 30°C.

Aromatic polyamides have excellent chemical stability, an operating range of 0°C to 35°C, a pH range between 4 and 11, and is resistant to biological attack. Polyamides are subject to degradation if exposed to chlorine.

Thin film composites have high chemical stability, high rejection and high flux rates at moderate pressures, a temperature range of 0°C to 40°C, and a pH range of 2–12. TFCs are susceptible to attack from chlorine and other oxidants.

MEMBRANE CONFIGURATIONS

Most membrane configurations are either thin film (spiral wound) or hollow fiber, but others are used as well. The spiral-wound configuration uses layers of membranes and supports, which are wrapped around a perforated permeate tube.[4] The spiral wound has good resistance to fouling because of relatively open feed channels, and because it is easy to clean, easy to replace, available in many varieties of membrane materials, and manufactured by a number of companies. The disadvantages of spiral wound configurations

---

[3]Brandt D, Leitner G, Leitner W. Reverse Osmosis Membranes State of the Art. In: Amjad, Zahid (Ph.D.), editor. Reverse Osmosis Membrane Technology, Water Chemistry, and Industrial Applications, New York: Van Nostrand Reinhold, 1993.
[4]Op. cit. footnote 3.

include low membrane surface area to volume ratio, possible concentration polarization, and difficulty in identification, isolation, and repairing individual elements in multiple-element tubes.

Hollow fiber membrane configurations are popular for a variety of reasons: (1) the hollow fiber configuration is formed by orienting the membrane fibers parallel within cylindrical pressure vessels, (2) pressure is applied to the fiber bundles from the outside, and the permeate flows to the interior of the bundle and through the length of the fiber, (3) the bundle of fibers has a high membrane surface area to volume ratio, and (4) it is easy to service and repair in the field. A number of manufacturers are making hollow fiber membrane bundles, and their popularity is increasing. The hollow fiber configuration is sensitive to the development of fouling by sediment and colloids and may be more difficult to clean than other types of systems.

Tubular configurations were some of the earliest RO devices ever introduced. Although not in use as often as hollow fiber and spiral wound modules, they still are used in applications with high levels of suspended solids, such as wastewater treatment. During operation with tubular modules, the high-pressure feed stream enters the tube, and the permeate passes through the membrane and supporting structure into a outer jacket where it is removed through permeate ports.

The advantages and disadvantages of the tubular configuration are as follows: The advantages include large flow passages, which permit the high flow velocities in the tubes, a low tendency to foul, an easiness to clean, and a high temperature stability; the disadvantages include low membrane surface area to volume ratio, high expense, and lower selection of material choices because of the demand for high tensile strength.

RO DESIGN CONSIDERATIONS

When designing a RO system, it is usually necessary to have both pretreatment and posttreatment of the water. The process should be viewed as a total system consisting of the following considerations: Feedwater supply, pretreatment, high pressure pumps, RO membranes, posttreatment, and end use. The following parameters are an overview of the considerations that one must take in designing a RO process.

Feedwater Supply Considerations:

Scale control, pH optimization, hardness, suspended solids content, metallic ions, organic chemical control and attack, biological inhibition of growth

Pressure Pumping:

System pressure, pump efficiency, pump flow rate and turndown rates, materials and corrosion

Membrane Considerations:

Configuration and type of membrane, material, salt rejection, chemical resistance, and recovery

Posttreatment:

pH adjustment, demineralization, degasification, disinfection, and storage.

Rejection is a common feature of RO systems. In working with seawater it is often common to have a rejection rate of about 4 gallons for every 1 gallon treated. The higher the salt concentration in the system, the harder it is to get a low salt concentration in the effluent.

Table 14.1 shows the common components for a pretreatment system for RO systems.

DESIGN PARAMETERS

Franklin Agardy goes through a complete list of design and operating parameters. Instead of rewriting these parameters myself, I prefer to quote his words here:

In large measure, both design and operational considerations of reverse osmosis systems are based on desalting experience. Nevertheless, the factors to be considered are equally valid in application to wastewater renovation. A brief discussion of these parameters will suffice to place into perspective each element in the overall performance.

Pressure: The water flux is a function of the pressure differential between the applied pressure and osmotic pressure across the membrane. The higher the applied pressure, the greater the flux. However, the pressure capability of the membrane is limited, and so the maximum pressure is generally taken to be 1,000 psig. Operating experience dictates pressures in the 400 to 600 psig range, with 600 psig normally being the design pressure.

Temperature: The water flux increases with increasing feedwater temperature. A standard of 70°F is generally assumed as an inlet design condition and temperatures up to 85°F are acceptable. However, temperatures in excess of 85°F and up to 100°F will accelerate membrane deterioration and cannot be tolerated for long intervals.

TABLE 14.1 Summary of Pretreatment Methods for RO from Reverse Osmosis Zahid Amjad, Editor, Van Nostrand, 1994

Problem	Pretreatment	Purpose	Problem Areas	Secondary Pretreatment	Purpose
Ca/Mg/bicarbonate scale	Ion exchange softening	Replaces cations with Na	High TDS causes slip. Maximum recommended is 800 mg/l, too expensive @ flow >9000 m^3/day	Sequestrant	Causes any tendency for slip to cause precipitations
	Lime softening	Removes Ca and HCO_3, precipitate Mg as $Mg(OH)_2$	Flow greater than 5000 m^3/day	Sequestrant or acid	Prevents postprecipitation
	Acid	Replaces bicarbonates with Cl or SO_4	Difficult and expensive	Sequestrant	Backup if acid dosing fails
	Sequestrant as primary conditioning	Not recommended because it delays formation of the precipitate	Other methods better suited more reliable		
Ca sulphate scale	Base exchange	See above			
	Sequestrant	See above			
	Raise temperature	Stabilizes and increases solubility of scale	Cost of energy		
Iron precipitates	Lime softening	See above	See above		
	Oxidize and filter	Removes precipitates	May cause problems with other materials like H_2S	Acid	Prevents further precipitates.
	Exclude oxidants	Keeps iron as filtrable Fe^{2+}	Not good for intermittense	Acid	Prevents precipitates

Acid		Keeps iron in solution	Need pH < 5	Sometimes acid works	Prevents postprecipitation
Colloids	Coagulate and filter Base exchange softening	Removes particles Stabilizes coagulation by changing Zeta potential of colloids and discourages precipitates	Not suitable for high TDS waters where total hardness after softening is >5 mg/l		
Bacteria	Sterilize with Cl_2 and filter	Remove with filtration	Hard waters need acid dose before treatment	Dose sodium metabisulfite	Removes Cl that could damage permeators
H_2S	Degas and add C_{12} Exclude oxidants	Removes H_2S Keeps H_2S in solution	Do not use if bacteria and present or if system has poor operation	See Bacteria above Need to degas permeant	H_2S!!!
Chlorine of oxidants	Dose with sodium metabisulfite or remove with activated carbon	Needs good maintenance and careful operation to avoid failure	Bisulfite can fail and cause harm to system. Carbon can breed bacteria	Need to monitor and back up system	If second dose is used, helps membrane life

Membrane Packing Density: This is an expression of the unit area of membrane which can be placed per unit volume of pressure vessel. The greater this factor, the greater will be the overall flow through the system. Typical values range from 30 to 500 sq. ft./cu./Ft. of pressure vessel.

Flux: Assuming a typical pressure of 600 psig, flux values range from 10 to 80 gpd/sq ft with 12 to 35 gpd/sq ft being common. This flux tends to decrease with length of run, and over a period of one to two years of operation might be reduced by 10 to 50%.

Recovery Factor: This consideration actually represents plant capacity and is generally in the range of 75–95 percent, with 80 percent being the practical maximum. At high recovery factors, there is a greater salt concentration in the process water as well as in the brine. At higher concentrations, salt precipitation on the membrane increases, causing a reduction in operational efficiency.

Salt Rejection: Salt rejection depends on the type and character of the selected membrane and the salt concentration gradient. Generally, rejection values of 85 to 99.5 percent are obtainable, with 95 percent being commonly used.

Membrane Life: Membrane life can be drastically shortened by undesired constituents in the feedwater, such as phenols, bacteria, and fungi, as well as high temperatures and high or low pHs. Generally, membranes will last up to two years with some loss in flux efficiency.

pH: Membranes consisting of cellulose acetate are subject to hydrolysis at high and low pHs. The optimum pH is approximately 4.7, with operating ranges between 4.5 to 5.5.

Turbidity: While reverse osmosis units can be used to remove turbidity from feedwaters, they operate best if little or no turbidity is applied to the membrane. Generally, it is felt that turbidity should not exceed one Jackson Turbidity Unit (JTU) and the feedwater should not contain particles larger than 25 microns.

Feedwater Stream Velocity: The hydraulics of reverse osmosis systems are such that velocities in the range of 0.04 to 2.5 fps are common. Plate and frame systems operate at higher velocity while hollow fine fiber units operate at the lower velocities. High velocities and turbulent flow are necessary to minimize concentration polarization at the membrane surface.

Power Utilization: Power requirements are generally associated with the system pumping capacity and operational pressures. Values range from 9 to 17 kWh/1,000 gal., with the lower figure taking into account some power recovery from the brine stream.

Pretreatment: The present development of membranes limits their direct application to feedwater having a total dissolved solids not exceeding

10,000 mg/l. Further, the presence of scale-forming constituents, such as calcium carbonate, calcium sulfate, oxides and hydroxides of iron, manganese and silicon, and possibly barium and strontium sulfates, zinc sulfide, and calcium phosphate, must be controlled by pretreatment or they will require subsequent removal from the membrane. These constituents can be controlled by pH adjustment, chemical removal, precipitation inhibition, and filtration. Organic debris and bacteria can be controlled by filtration, carbon pretreatment and chlorination. Oil and grease must also be removed to prevent coating and fouling of membranes.

Cleaning: Recognizing that under continuous use membranes will foul, provisions must be made for mechanical and/or chemical cleaning. Methods reviewed include periodic depressurizations, high velocity water flushing, flushing with air-water mixtures, backwashing, cleaning with enzyme detergents, ethylene diamine tetra-acetic acid and sodium perborate. The control of pH during cleaning operations must be maintained to prevent membrane hydroxysis. Approximately 1.0 to 1.5 % of the process water goes to waste as a part of the cleaning operation with the cleaning cycle being every 24 to 48 hours.

A summary of operational parameters is given in Table 14.2.

TABLE 14.2   Summary of Operational Parameters for RO Systems

Parameter	Range	Typical
Pressure (psig)	400–1000	600
Temperature (°F)	60–100	70
Packing density (ft^2/ft^3)	50–500	—
Flux (gallons/day/ft^2)	10–80	12–35
Recovery factor (%)	75–95	80
Rejection factor (%)	85–99.5	95
Membrane life (years)	—	2
pH	3–8	4.5
Turbidity	—	1 JTU
Feedwater velocity (ft/s)	0.04–2.5	—
Power utilization	9–17 kWh/1000 gallons	

15

CARBON ADSORPTION

Breakthrough curves
The Freundlich and the Langmuir equations
Carbon adsorption physical coefficients and economics
PACT™ process

Adsorption is a basic process in Chemical Engineering. It is a process we all have studied in undergraduate classes. The review on this section will touch only the highlights as a refresher.

BREAKTHROUGH CURVES

In normal adsorption, a typical breakthrough curve is shown as follows (Fig. 15.1):

Factors Affecting Adsorption:

1. Particle diameter (inversely with absorbent particle size [inverse of surface area]).
2. Adsorbate concentration (directly varies).
3. Temperature (direct variation).
4. Molecular weight (Generally an inverse variation depending upon the compound weight and configuration of pore diffusion controls).
5. pH (inverse with pH due to surface charge).

Practical Wastewater Treatment, by David L. Russell
Copyright © 2006 John Wiley & Sons, Inc.

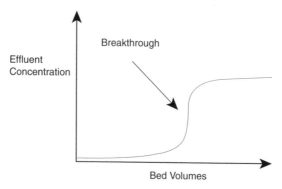

FIGURE 15.1   Breakthrough curve for carbon adsorption.

6. Individual properties of solute and carbon are difficult to compare.

7. Iodine number.

THE FREUNDLICH AND THE LANGMUIR EQUATIONS

The basic equations for adsorption are the Freundlich and the Langmuir equations.

Langmuir equation: $y/m = Kc/(1 + K_1 c)$, where K and K_1 are determined constants, and y/m is the amount of material adsorbed per unit weight of sorbent.

Freundlich equation: $y/m = Kc^{1/n}$, where c is the concentration of material in solution at equilibrium, and K and c are experimentally determined constants. Depending upon the application, one can use either equation. The most common is the Freundlich equation when it is expressed as:

$\log(y/m) = \log K + (1/n)\log(c)$, which plots as a linear form on log paper and makes the determination of the constants relatively easy. The Langmuir equation when expressed in a convenient form looks like the following:

$c/(y/m) = 1/K + K_1/Kc$, and the variation between $c/(y/m)$ and c is linear. Under certain conditions, the Langmuir equation may create a better fit for the data than the Freundlich equation.

Although the manufacturer's curves are fine for initial design, one should use experimentally determined constants on a specific wastewater because of interferences and variables in the wastewater.

CARBON ADSORPTION PHYSICAL COEFFICIENTS AND ECONOMICS

An excellent source for information on adsorption isotherms for toxic organics was prepared by the EPA and is still available from the NTIS.

"Carbon Adsorption Isotherms for Toxic Organics", EPA 600/8-800-023 April 1980. Summary data from that source are presented later in this chapter. The values were determined in an "artificial wastewater", which contained approximately 200 mg/l of alkalinity, sodium, calcium, and other ions commonly found in wastewater.

Other Considerations

Carbon Regeneration   The carbon can be regenerated and reused. Most of the times, the carbon can be regenerated by thermal oxidation or steam oxidation. However, the cost of carbon is significant. The following generally applies to situations in the United States.

- The carbon is expensive. Virgin carbon can be $1–$1.40/lb. - ($2.2–$3/kg). Regeneration is about 80% of the cost of new carbon.
- If certain compounds are removed from the waste, the carbon could be classified as a hazardous waste, requiring special treatment. Hazardous waste disposal costs can be $500/ton (U.S.) plus $3/ mile for shipment to a hazardous waste landfill.

PACT^{TM} PROCESS

PACT^{TM} was originally developed by DuPont, but is now owned by Zimpro. (See the following link for description: http://p2library.nfesc. navy.mil/P2_Opportunity_Handbook/9-IV-4.html). The process adds powdered activated carbon to the wastewater treatment tank. It is used where there are biologically resistant organics or toxics in the wastewater, and it provides a combination of carbon pretreatment and increased retention time that enables the bacterial population to acclimatize and degrade the organics. It is also useful in reducing some metal concentrations. Carbon dosages vary considerably with the organics, and the activated carbon adds solids to the clarifier and can add substantially greater quantities of abrasives to the clarifier underflow. Generally, the concentration maintained in the aeration basin is under several hundred milligrams per liter.

The PACT^{TM} process uses a high temperature, high-pressure water treatment with a copper catalyst. The process operates at temperatures up to about 600 psi and temperatures up to 300°F–400°F as an oxidizing system. This process is called "wet oxidation" and has been marketed successfully for a number of years by "Zimpro," now a division of U.S. Filter Corp.

Overview of Zimpro Technology

Wet oxidation is the oxidation of soluble or suspended oxidizable components in an aqueous environment using oxygen (air) as the oxidizing agent. The oxidation reactions occur at elevated temperatures and pressures.

Wet Air Regeneration for PACT Systems   Wet air regeneration (the Zimpro process) is a liquid phase reaction in water using dissolved oxygen to oxidize sorbed contaminants and biosolids in a spent carbon slurry, while simultaneously regenerating the powdered activated carbon.

Regeneration is conducted at moderate temperatures of 400–500°F (205–260°C) and at pressures from 700–1000 psig (50–70 bar). The process converts organic contaminants to CO_2, water, and biodegradable short chain organic acids; sorbed inorganic constituents such as heavy metals are converted to stable, nonleaching forms that can be separated from the regenerated carbon, if necessary.

The system is claimed to be more cost-effective and energy efficient than that of furnace technology for regeneration. Regeneration is done in a slurry without NO_x, SO_x, or particulate air emission problems.[1] According to Zimpro, the operating cost for PACT can range between $0.50 and $1.00/1000 gallons treated ($0.13–$0.30/M³).

Simplified, general wet oxidation flow diagram and coefficient of carbon adsorptions are shown in Figure 15.2 and Table 15.1, respectively.

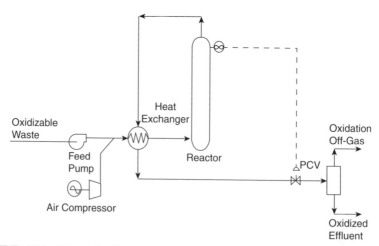

FIGURE 15.2 Schematic diagram of Zimpro wet oxidation process for treating and regenerating powdered activated carbon (PACT process).

[1]www.zimpro.com.

TABLE 15.1 Summary of Carbon Adsorption Capacities

Compound	Adsorption[a] Capacity, mg/g	Compound	Adsorption[a] Capacity, mg/g
bis(2-Ethylhexyl) phthalate	11,300	Phenanthrene	215
		Dimethylphenylcarbinol*	210
Butylbenzyl phthalate	1,520	4-Anhinobiphenyl	200
Heptachlor	1,220	β-Naphthol*	200
Heptachlor epoxide	1,038	α-Endosulfan	194
Endosulfan sulfate	686	Acenaphthene	190
Endrin	666	4,4'-Methylene-bis- (2-chloroaniline)	190
Fluoranthene	664		
Aldrin	651	Benzo(k)fluoranthene	181
PCB-1232	630	Acridine orange*	180
β-Endosulfan	615	α-Naphthol	180
Dieldrin	606	4,6-Dinitro-o-cresol	169
Hexachlorobenzene	450	α-Naphthylamine	160
Anthracene	376	2,4-Oichlorophenol	157
4-Nitrobiphenyl	370	1,2,4-Trichlorobenzene	157
Fluorene	330	2,4,6-Trichlorophenol	155
DOT	322	β-Naphthylamine	150
2-Acetylaminofluorene	318	Pentachlorophenol	150
α-BHC	303	2,4-Dinitrotoluene	146
Anethole*	300	2,6-Dinitrotoluene	145
3,3-Dichlorobenzidiene	300	4-Bromophenyl phenyl ether	144
2-Chloronaphthalene	280	p-Nitroaniline*	140
Phenylmercuric Acetate	270	1,l-Diphenylhydrazine	135
Hexachiorobutadiene	258	Naphthalene	132
γ-BHC (lindane)	256	1-Chloro-2-nitrobenzene	130
p-Nonylphenol	250	1,2-Dichlorobenzene	129
4-Dimethylaminoazobenzene	249	p-Chlorometacresol	124
Chlordane	245		
PCB-1221	242	1,4-Dichlorobenzene	121
DDE	232	Benzothiazole*	120
Acridine yellow*	230	Diphenylamine	120
Benzidine dihydrochloride	220	Guanine*	120
β-BHC	220		
N-Butylphthalate	220	Styrene	120
N-Nitrosodiphenylamine	220	1,3-Dichlorobenzene	118
2-Nitrophenol	99	Acenaphthylene	115
Dimethyl phthalate	97	4-Chlorophenyl phenyl ether	111
Hexachloroethane	97	Diethyl phthalate	110
Chlorobenzene	91	Bromoform	20
p-Xylene	85	Carbon tetrachloride bis- (2-Chloroethoxy) methane	11
2,4-Dimethylphenol	78		11
4-Nitrophenol	76	Uracil*	11
Acetophenone	74	Benzo(ghi)perylene	11
1,2,3,4-Tetrahydro- naphthalene	74	1,1,2,2—Tetrachloroethane	11
		1,2-Dichloropropene	8.20

(*Continued*)

TABLE 15.1   (*Continued*)

Compound	Adsorption[a] Capacity, mg/g	Compound	Adsorption[a] Capacity, mg/g
Adenine*	71	Dichlorobromomethane	7.90
Dibenzo(*a,h*)anthracene	69	Cyclohexanone*	6.20
Nitrobenzene	68	1,2-Dichloropropane	5.90
3,4-Benzofluoranthene	57	1,1,2-Trichloroethane	5.80
1,2-Dibromo-3-chloro-		Trichlorofluoromethane	5.60
propane	53	5-Fluorouracil*	5.50
Ethylbenzene	53	1,1-Dichloroethylene	4.90
2-Chlorophenol	51	Dibromochloromethane	4.80
Tetrachloroethene	51	2-Chloroethyl vinyl	
o-Anisidine*	50	ether	3.90
5 Bromouracil	44	1,2-Dichloroethane	3.60
		1,2-*trans*-Dichloroethene	3.10
Benzo(*a*)pyrene	34	Chloroform	2.60
2,4-Dinitrophenol	33	1,1,1-Trichloroethane	2.50
Isophorone	32	1,1-Dichloroethane	1.80
Trichloroethene	28	Acrylonitrile	1.40
Thymine*	27	Methylene chloride	1.30
Toluene	26	Acrolein	1.20
5-Chlorouracil 1*	25	Cytosine*	1.10
N-Nitrosodi-*n*-propylamine	24	Benzene	1.00
bis(2-Chloroisopropyl)		Ethyene-diamine-D33tetra-	
ether	24	acetic acid	0.86
Phenol	21	Benzoic acid	0.76
		Chloroethane	0.59
		N-Dimethylnitrosoanaline	6.8 * E-05

(a) Note: Distilled water used with the following ion addition (mg/l)

Na	92	PO_4	10
K	12.6	SO_4	100
Ca	100	Cl	177
Mg	25.3	Alkalinity	200

16

ION EXCHANGE

INTRODUCTION

The ion exchange (IX) process occurs when ions that are held to functional groups on a solid surface by electrostatic forces are exchanged for ions of a like charge in a solution in which the solid is immersed. The solid is called a resin.

IX can be performed either in batch processes or in columns. Batch systems are less complex than the columnar system; however, they are also inefficient. For this reason, most IX processes are performed in a column.

RESINS

There are both synthetic and natural resins. Natural resins have been called zeolites, greensands, clinoptilolites, and natrolites. Soils and peat materials also have some smaller amount of ion exchange capacity. However, with few exceptions, most of the resins used today are synthetic. They are made of a polymer matrix with soluble ionic functional groups attached to the polymer chains. When the resin is used up, a concentrated solution of the charged functional group can be applied to regenerate the resin. Because the resin is expensive, in most cases the regeneration is an economical must.

Physical Characteristics

Resins may be in the form of either a gel or a macroporous resin. Macroporous resin exchange sites may be lower on a volume basis. Although the quantity of regeneration may be greater, resins are more resistant to thermal and osmotic shock and oxidation, are less susceptible to fouling, and have a very long life. But, each quality has its own cost, subsequently making resins a lot more expensive.

Chemical Structure

Strong acid resins contain sulphonic acid groups as the exchange sites. They have a regeneration efficiency of 30% to 50%. That regeneration is usually done with strong acids such as H_2SO_4 or HCl.

Weak acid resins contain carboxylic acid groups as the functional species. They are extremely stable thermally and can be regenerated with any acid that is stronger than the functional group. Regeneration is nearly 100%. However, they must only be used in water with a pH greater than 7.

Strong base resins usually contain quaternary amine groups as the functional species. Regeneration is usually done with NaOH and has an efficiency of 30% to 50%.

Weak base resins can contain tertiary $(-NR_2)$, secondary $(-NHR)$, or primary $(-NH_2)$ amino groups, or a mixture of them as the functional species. The water must have a pH less than 7. They can be regenerated by NaOH, Na_2CO_3, or NH_4OH at nearly 100% efficiency.

Chelating resins are developed to be more selective toward certain ions compared with others. They can be regenerated under acidic conditions because these are weakly acidic. Many of these resins are imidodiacetic acid groups attached to some cross-linked polystyrene. Some ion exchangers containing specific groups that are selective for particular ions are shown in Table 16.1.

TABLE 16.1   Selective Chealating Resins in Ion Exchange

Type of Ion	Specific Compound
Nickel, Mercury, other select heavy metals	Thiol (Azko Chemicals)
Copper	Amidoxime (Duolite)

SELECTIVITY

Selectivity is a property of an ion exchange medium; it represents the preferential activity the medium has for different ions. This is also often related to valence. It shows that compounds of higher valence, either positive or negative, will be preferentially exchanged.

At low concentrations and room temperature, polyvalent ions get preference over monovalent ions.

Generally, ion exchangers prefer counter ions, which

have a higher valence;
are smaller in equivalent volume;
have greater polarizability;
interact more strongly with the fixed ionic group of the matrix; and
participate least in combining into complex formulations with the co-ions.

SELECTIVITY COEFFICIENT

Ion exchange reactions are stoichiometric and reversible. They are of the following type:

$$R\text{-}A^+ + B^+ \leftrightarrow R\text{-}B^+ + A^+$$

where R is the resin, A^+ is the functional ion, and B^+ is the ion originally found in the solution. The degree to which the exchange occurs depends upon the selectivity of the resin for the exchanged ion.

The selectivity coefficient K is defined as the relative distribution of ions when a charged resin is made to contact with different, but similarly charged, ions.

$$K = \frac{[B^+]\,\text{in resin}}{[A^+]\,\text{in resin}} \times \frac{[A^+]\,\text{in solution}}{[B^+]\,\text{in solution}}$$

The magnitude of K represents the relative preference to absorb $[B^+]$ as compared with $[A^+]$; the greater the magnitude of K, the greater the preference for the ion by the exchanger. Table 16.2 shows selectivities of IX resins in the order of decreasing preference.

When there is a high affinity for the ion to be exchanged there is a sharp breakthrough curve, a shorter IX column, and a greater flow rate. However, a higher regenerant concentration is required.

TABLE 16.2   Ion Preference and Affinity for Selected Compounds

Strong Acid Cation Exchanger	Strong Base Anion Exchanger	Weak Acid Cation Exchanger	Weak Base Anion Exchanger	Weak Acid Chelate Exchanger
Barium (2+)	Iodide (1−)	Hydrogen (1+)	Hydroxide (1−)	Copper (2+)
Lead (2+)	Nitrate (1−)	Copper (2+)	Sulfate (2−)	Iron (2+)
Mercury (2+)	Bisulfite (1−)	Cobalt (2+)	Chromate (2−)	Nickel (2+)
Copper (1+)	Chloride (1−)	Nickel (2+)	Phosphate (2−)	Lead (2+)
Calcium (2+)	Cyanide (1−)	Calcium (2+)	Chloride (1−)	Manganese (2+)
Nickel (2+)	Bicarbonate (1−)	Magnesium (2+)		Calcium (2+)
Cadmium (2+)	Hydroxide (1−)	Sodium (1+)		Magnesium (2+)
Copper (2+)	Fluoride (1−)			Sodium (1+)
Cobalt (2+)	Sulfate (2−)			
Zinc (2+)				
Cesium (1+)				
Iron (2+)				
Magnesium (2+)				
Potassium (1+)				
Manganese (2+)				
Ammonia (1+)				
Sodium (1+)				
Hydrogen (1+)				
Lithium (1+)				

DESIGN CONSIDERATIONS

Pretreatment

Pretreatment for solids removal is often required. What is needed here is common sense. Anything that can attack the resin, including acids, organics, and even some solids, should be avoided. Resins also do not like strong acids or bases. Many use salt for regeneration. Iron and calcium can blind the resins, as can certain types of silicates.

Note that when an exchange takes place, it does not necessarily affect other ions in the system. Thus,

$$Ca(HCO_3)_2 + Na_2AZ \leftrightarrow CaZ + 2(NaHCO_3),$$ so calcium bicarbonate hardness is exchanged for sodium bicarbonate. Note that nothing happens to the bicarbonate.

It is possible to exchange in a mixed bed ion exchange unit. That allows both cations and anions to be exchanged at the same time. If demineralized water is desired, then hydrogen and hydroxide ion exchange resins should be utilized.

On the basis of old water treatment measurements of hardness, being measured as grains/gallon (U.S.), many resins have exchange capacity expressed in grains/ft^3.

1 grain = 1 lb/7000 or
1 grain/ft^3 = 2.884 g/m^3, and
1 grain/gallon = 17.118 mg/l

Ion exchange capacity is often expressed in terms of milliequivalents of $CaCO_3$. The milliequivalents are calculated on the basis of an assumed molecular weight of 100.00 rather than 100.08.

Backwash volume is often 8–12% of throughput volume. The backwash will contain the regenerant plus the material removed and may be a hazardous waste.

Demineralized water is often aggressive water. The selection of materials is important, and plastic or glass pipes may be required for certain types of backwash and tank linings.

During backwash, the bed can be handled as either upflow or downflow. Bed suspension is not always necessary but it is recommended on larger units.

For more help and information on sizing ion exchange resins, go to the Rohm and Haas Web sites listed below. They have a sizing calculator, which is very handy. The Osmonics website is equally informative.

http://www.osmonics.com/products/page838.htm
http://www.rohmhaas.com/ionexchange/fr_resins.htm
http://www.rohmhaas.com/ionexchange/fr_special.htm

17

DISSOLVED AIR FLOTATION AND TECHNIQUES

Design basics for DAF
Operating parameters
Electroflotation
Electrocoagulation

DESIGN BASICS FOR DAF

Dissolved air flotation (DAF) and flotation in general is mostly dependent upon the solubility of nitrogen and oxygen in water.

As we saw earlier, the normal saturation value for oxygen in water is about 11 ppm at common environmental conditions. Flotation is removal of suspended solids by the process of reverse stokes settling and coagulation. Some flotation units use induced air at low pressures, while others use compressed air. The mining industry uses aerators to beat the air into the water (much like a kitchen mixer) for separation of the ores in a process called beneficiation. The beneficiation process usually handles large quantities of solids, which have a greater density than those generally handled by most environmental flotation processing plants. Many environmental flotation facilities are used for removing grease, oils, fats, and low-density solids from the wastewaters. Some commercial bakeries, dairies, fish processing, and poultry plants use DAF to remove everything from fats to blood from the wastewater.

In "conventional" dissolved air flotation, a part of the flow is pressurized between 40 and 100 psig (2.72–6.8 atm). At those pressures, nitrogen and

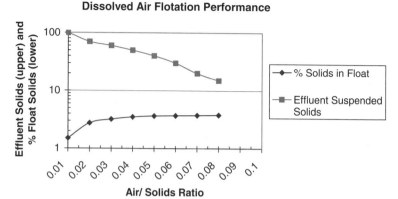

FIGURE 17.1   Dissolved air flotation system performance.

oxygen are substantially more soluble in water than at atmospheric pressure. For example, the release of nitrogen from decompression at 40 psi is about 211 cm^3/l, and that of oxygen is about half of that value. So, overall from the release of pressure at 40 psig, we can generate about 320 ml of gas per liter of water pressurized.

Most environmental solids have a density less than 2 gm/cm^3. Silica has a density of 2.65 gm/cm^3. In Chapter 7 we discussed reduction in apparent specific gravity by particle agglomeration. If air (density 1.28×10^{-3} gm/cm^3) can be made to adhere to a sand or silt particle, it does not take many bubbles to make even sand "float".

This was calculated from the Henry's law constants we used in Chapter 1 (see Table 1.2 of Chapter 1).

The inverse of the Henry's law constant, multiplied by the partial pressure of the gas above the solution, is the molar solubility of the gas. Thus oxygen at 1 atm would have a molar solubility of (1/756.7) mol/dm^3 or 1.32 mmol/dm^3.

The key to measurement is the air/solids ratio. Typical curves are shown in Figure 17.1.[1]

The design of a DAF unit is relatively straightforward. A typical design configuration is shown in Figure 17.2.

OPERATING PARAMETERS

Liquid overflow rate is somewhat higher than that of a clarifier; 0.7–2.7 L/M^2/S (2–5 gallons/min/ft^2 [this is a vertical velocity or overflow rate on a

---

[1]Eckenfelder, Thackston. New Concepts in Wastewater Treatment. New York: Jenkins Press, 1974.

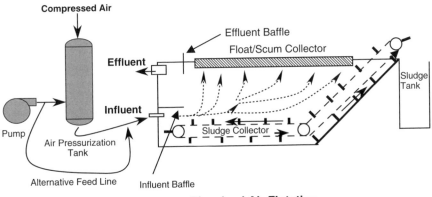

**Dissolved Air Flotation
Unit**

FIGURE 17.2   Dissolved air flotation system configuration.

par with filtration rates]) or higher if laboratory tests indicate. Air/solids ratios from 0.01 to 0.2 have been used in design, but as a practical matter 0.03–0.05 air/solids ratios will give the best suspended solids removals. However, laboratory and pilot tests must be used to determine the best values. Detention times in the system can vary from about 15 min to over 1 h. Side water depths of the tanks are between 1.3 m and 3 m (5–10 ft).

The recycle ratio for water ranges from 10% to 100%, with general values from 20% to 60% depending upon the solids concentration and chemical addition.

With proper coagulant dosage and emulsion breakers as required, the solids removals can easily approach 95–99% of the suspended solids, and depending upon the chemistry, 50% or more of the total dissolved solids. When de-emulsifying oily wastewaters, DAF units have been known to produce an effluent with less than 5 mg/l total oil, but 15 mg/l is much more reliable and attainable, even with influent concentrations of 1000–16,000 mg/l oil.

Theory and Design

One theory of removal by DAF depends upon the collision theory and the work of Tambo and Wantanabe (1968).[2] Tambo and his co-worker developed a theory indicating that population of particles with air bubbles attached depends upon kinetic factors, and the concentration of the particles, the

[2]Tambo N, Wantanabe Y. A Kinetic Study Dissolved Air Flotation. Tokyo: World Congress of Chemical Engineering, 1968, pp. 200–203.

concentration of the air bubbles, and the collision between the two. The result is a rather messy differential equation involving a first-order differential equation and a rather messy integration of the form of:

$$dN/dt = -kN(a_m N_{bo} - a_{m-1} N_{bo-1})$$

where N is the number of air bubbles; N_{bo} is the concentration of particles without air bubbles; k is a kinetic coefficient; and a is an attachment factor.

The important thing that has come up with this work was the realization that the kinetic coefficient is equal to the velocity gradient times the cube of the sum of the bubble and floc diameters. The bubbles have a fixed size from 40 to 100 microns, and the collision rate increases with floc size.

It is far easier to use a model that is analogous to a filter contact model. This model is the "Whitewater" model, because it describes the condition of the water in the saturation zone where the air is released into the water.

The basic assumption in the model is that:

$$\eta_t = \eta_d + \eta_i + \eta_s$$

where η_t = total collisions; η_d = Brownian diffusion; η_i = interception; and η_s = differential settling (particles relative to bubbles).

Furthermore, there is an attachment ratio or "a", which is further applied to represent the missed particles.

Rather than going through and giving the entire theory, we will break down the significant equations:

(1) Bubble Mass

$$C_b = (C_r - C_n)r/(1 + r)$$

where C_b = Mass concentration of air released; C_r = Mass concentration of air in the recycle flow (mg/l); C_n = Mass concentration of air in the floc tank effluent (mg/l); r = recycle ratio (decimal).

(2) Particle Bubble Rise

$$V_{pb} = \text{Stokes law} = g(\rho_p - \rho_{pb})d_{pb}2/18\mu$$

where V_{pb} = rise velocity of the particle + bubble (m/h); g = gravity; μ = viscosity; and ρ = density of the particle or the particle plus the bubble, respectively.

(3) Bubble Volume Concentration

$$\Phi_b = C_b/\rho_{air}$$

where C_b is given in (1), and ρ_{air} is the density of air saturated with water vapor.

(4) **Bubble Number Concentration**

$$N_b = 6\Phi_b/(\pi d_b^3)$$

Φ_b is from (3), and d_b is the mean bubble diameter in microns.

Ranges of Data

Particles per milliliter range from 1000 to 10,000 for drinking water applications, and on the basis of TSS loadings alone, for many industrial applications the values for industrial wastes could be 100,000–300,000 per milliliter.

Bubble diameters depend upon saturator pressure and recycle rates. For recycle rates between 6% and 15%, the general estimate of bubble diameters is about 40 microns median size, and the number of bubbles for low solids content water is between 10^5 and 2×10^5 bubbles per milliliter and the estimated bubble to particle ratio of approximately 200:1.

For various types of waste streams, the amount of air is often independent of the suspended solids in the system, unless the TSS is more than 1000 mg/l. For surface waters low in solids, the approximate range of air/solids ratio is about 380 ml of air per gram of solids. For sludge thickening applications, the air requirement is between 15 and 30 ml/g.[3]

ELECTROFLOTATION

Electroflotation is a much-overlooked technology. It is accomplished by disassociation of water by electricity either in an atmospheric tank or in a pressure tank. The atmospheric tank is the most common application. The equipment is still used in the oil industry, especially in locations where the conductivity of the oily water is above 1000 micromohs.

The basic reactions are the ones for the disassociation of water:

$$2H_2O + 4e \leftrightarrow 2H_2 + O_2$$

For every 4 Columbs of electricity one gets 2 moles of hydrogen and 1 mole of oxygen released into the water. If there are salts such as chlorides, they will also disassociate into chlorine, at some reduced efficiency, but at no electrical penalty.

---

[3]See Chapter 7 in Water Quality and Treatment by the American Water Works Association, 5th Ed., New York: McGraw Hill.

The use of this system generates a cloud of microbubbles that are far more gentle on flocs than those generated by many of the existing systems, which utilize pressurized air/water systems. Also, the flotation is continuous while the power is turned on. Regulation can be by control of the power supplied to the unit, and one can obtain greater or smaller amounts of flotation by adjusting the "gas/solids" ratio as opposed to an air/solids ratio. In this case, the gas/solids ratio is appropriate because the system does not use air, but the principal mechanism of flotation is excess hydrogen and oxygen generated at the level of the electrode and also allowed to bubble to the surface.

The power generation requirements are modest and can compete successfully with other types of flotation, especially where high recycle rates are employed. The applied voltage is dependent upon the conductivity of the water, but seldom above 12 V. The amperage is significant and can be several hundred amperes. One system used a maximum voltage of 12 V, and about 300 A, or about 3.6 KVA to treat up to 100 gallons per min in a flotation system basin of approximately 6.2 m^2, or 64 ft^2. The bubbles are small, (between 10 and 20 microns, and many smaller) but there are many more of them and they have a lower density than that of air.

One of the few disadvantages of the flotation system is the fact that some of the byproducts of the flotation are hydrogen and oxygen, which are trapped in the foam, and perhaps some free chlorine if chlorides are present in the water. If the foam generated is viscous, it can trap and retain the hydrogen and oxygen together, creating the potential for a small hydrogen explosion, which is limited because of the limited amount of foam and the presence of water in the foam.

When ignited, it can sound like a firecracker going off, but, at the same time, one can see how this would upset the safety of people and cause concern. Given a well ventilated room, and/or a vacuum system to collect and collapse the foam, the hazard is minimal.

One type of electrode configuration is shown in Figure 17.3.

TYPICAL FLOTATION GRID STRUCTURE

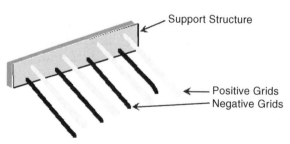

FIGURE 17.3 Electroflotation system grid configuration.

These systems have been used, in the mid- and late 1970s, to perform oil removal, predominantly in the meat packing industry. The U.S. Air Force rediscovered the technology through a technology exchange program with Russia in the early 1990s and tried to commercialize it for treating oily wastewaters from machine shops. Some of the systems are still in use in the oil industry working principally on brines for oil separation. However, most of the systems are of a marginal design, which limits their effectiveness and performance.

The significant problems with this type of system were found to be the electrode materials of construction. Those problems can be solved by selection of materials of electrode construction including high silicon iron, carbon, stainless steel, and titanium. There are other uses of the technology, also called electrocoagulation.

Electroflotation Theory and Design

The theory discusses the depth, length, and other parameters involved in sizing the electrodes and determining the electrical losses in the system.

The power consumption in passing an electric current between two long bars or rods can be found to be a function of the current passed through the circuit, the diameter of the rods, the length of the rods, the temperature of the solution, the depth of immersion of the electrodes, the resistivities of the electrode materials, the gap between the rods, and the molality of the electrolyte.

The power requirement for the system is the product of the current and the total circuit voltage, where the total circuit voltage will be the sum of the decomposition voltage of the electrolyte, the anode overvoltage, the cathode overvoltage, the ohmic drop through the electrolyte, and the ohmic drop through the electrodes and buss works.

$$\text{Power} = I\,E_{\text{Total}}$$

$$E_{\text{Total}} = E_{\text{Decomposition}} + \eta_a + \eta_c + \eta_\Omega + \eta_{hw}$$

$$E_{\text{Decomposition}} = E_{\text{Rev},T,P=1} + \frac{RT}{nF}\ln\frac{a_{H_2}(a_{O_2})^{1/2}}{a_{H_2O}}$$

$$E_{\text{Rev},T,P=1} = 1.5184 - 1.5423 \times 10^{-3}T + 9.524 \times 10^{-5}T\ln T$$

$$+\, 9.84 \times 10^{-8}T^2$$

The temperature ranges from 298 K to 523 K, and the pressures are expressed in atmospheres.

$$E_{\text{Decomposition}} = E_{\text{Rev},T,P=1} + 4.309 \times 10^{-5} T \ln \frac{(P - P_{H_2O})^{1.5} P_{H_2O}^o}{P_{H_2O}}$$

$$\ln P_{H_2O}^o = 37.043 - \frac{6275.7}{T} - 3.4159 \ln T$$

$$\ln P_{H_2O} = 0.016214 - 0.13802m + 0.19330m^{1/2} + 1.0239 \ln P_{H_2O}^o$$

where m is the molality of electrolyte in moles per kilogram of solvent.

$$P = P_{\text{atm}} + \frac{\text{sg}_{\text{solution}}(\text{depth})}{33.9}$$

(sg denotes the specific gravity; and the depth is in feet)

$$\eta_{hw} = \left(\frac{I\,L\rho}{3\,A}\right)_{\text{anode}} + \left(\frac{I\,L\rho}{3\,A}\right)_{\text{cathode}}$$

where L is the length of the electrode, and A is the cross-sectional area, $\pi D^2/4$, and ρ is the resistivity of the electrode material.

$$\eta_c = 2.303 \frac{RT}{F\alpha} \log \frac{i}{i_o}$$

where $R/F = 4.309 \times 10^{-5}$, α is 0.5, and -log i_o is ~6 for iron cathode, and i is current density in A/cm^2.

$$\eta_a = 0.37 \log \frac{i}{1.7 \times 10^{-5}}$$

The equation is for an iron anode; for both anode and cathode over voltages, the electrode material plays the largest influence.

$$\eta_\Omega = iL_{\text{gap}}\rho$$

where i is the current density in A/cm^2, L_{gap} is the distance between electrodes in cm, and ρ is the resistivity of the electrolyte.

Note: current density is given by current divided by active surface area of electrode, or $(\pi D^2/4)L$ (diameter and length of electrode).[4]

ELECTROCOAGULATION

The difference between electrocoagulation and electroflotation is only in the design of the electrode. In electroflotation, one uses high silica iron or other materials, which are essentially resistant to erosion by impressed current. In electrocoagulation, the electrode is designed to be sacrificial. The principles are the same for both, but as the electrode is eroded, the voltages will have to be higher to compensate for the increased spaces between the electrodes.

This technology is the equivalent of adding iron or alum directly to the water without the anion half of the compound. The ions are placed in water as hydroxides rather than as a sulfate, chloride, or other ion. In some instances, this technology has been coupled with flotation for enhanced removals. These systems were being manufactured by Kwire, a Japanese company, who apparently is no longer in business. (http://www.kwire.com/watertr.htm)

The following report from EPA covers the Electrocoagulation subject: http://www.epa.gov/ORD/SITE/reports/540r96502.pdf. See the site with the pdf file. This site also has specific information that illustrates the discussion: http://www.raintech.com. As of this writing, the Raintech Web site was no longer available on the World Wide Web, the site was registered in China and it expired in 2001. The information contained in the description taken from the Web site is largely accurate, but as with all manufacturer's claims, it should be examined carefully, as it may be selective and limited to a single instance.

The following texts have been taken from the Raintech Web site:

Electrocoagulation Vs. Chemical Coagulation

Because Electrocoagulation (EC) utilizes methods that precipitate out large quantities of contaminants in one operation, the technology is the distinct economical and environmental choice for industrial, commercial and municipal waste treatment. The capital and operating costs are usually significantly less than chemical coagulation. It is not unusual to recover capital costs in less than one year. (Editorial Note: All manufacturer's claims as to

---

[4]Comprehensive Treatise of Electrochemistry, Vol. 2,6, Plenum Press, 1981 and Electrochemical Cell Design, Plenum Press, 1984.

savings and payback need to be closely examined as the claims are often made upon unrealistic assumptions.)

For example a 5 GPM system contrasts the advantages of Electrocoagulation with a typical chemical coagulation system. This system was designed with the following requirements:

Reduce Ni from 8.74 to < 3 mg/l
Reduce Zn from 28.0 to < 3 mg/l
Reduce TSS from 657 to < 350 mg/l
Reduce Oil and Grease from 27 to < 15 mg/l
Reduce Phosphorus from 158.75 to < 10 mg/l
Process flow rate of 5 GPM (1,500,000 GPY)

{Editorial Note: These same results or better can be achieved by careful addition of either alum or ferric chloride and polymer in a standard flotation system or a standard precipitation system.}

The estimated yearly operating cost saving using Electrocoagulation in place of chemical coagulation is $43,500.00 per year. This does not include labor, sludge transportation or disposal costs.

A second example is a system with requirements to:

Reduce Ni from 25 to < 2.38 mg/l
Reduce Cr from 210 to < 1.71 mg/l
Flow rate of 100 GPM (30,000,000 GPY)

Operating cost:

Chemical Coagulation vs. Electrocoagulation per 1,000 gal
$14.18   vs. $1.69

Yearly Chemical costs:

$425,400.00   vs. $50,700.00

Electrocoagulation uses electricity to precipitate the dissolved and suspended solids. The total dissolved solids in the liquid usually decrease by 27 to 60 percent...

{Note: in Table 17.1, there is no comparison with DAF}

TABLE 17.1   Comparison Between Electroflotation and Sedimentation/ Precipitation

Parameter	Percentage Removal by Electroflotation (%)	Percentage Removal by Sedimentation/ Precipitation (%)
TSS	95–99	80–90
BOD	50–98	50–80
Bacteria	95–99.9	80–90

The rest of the information on the Web site was largely theoretical calculations, which further highlighted the value of their system without substantiation. Since the Web site is no longer available, it serves as an example of the lack of commercialization of this type of technology.

18

COAGULATION, FLOCCULATION, AND CHEMICAL TREATMENT

Introduction
Flocculation and mixing
Practice
Modeling

INTRODUCTION

Coagulation is all about bringing things together. It is joined with flocculation and chemical treatment because all the processes are necessary for chemical treatment and precipitation.

Coagulation is defined as destabilization by particle charge neutralization and initial aggregation of colloids.

Flocculation is agglomeration of coagulated colloidal and finely divided suspended material either by physical mixing or by chemical coagulant aids.

Chemical treatment is what we do to make coagulation and flocculation happen by adjusting the chemical charges on contaminants through the process of adding chemicals.

The most effective coagulant aids are divalent and trivalent metallic ions, usually iron and aluminum, but can include calcium, magnesium, and manganese. The other things that can be used as coagulant aids are polymers and sols.

Practical Wastewater Treatment, by David L. Russell
Copyright © 2006 John Wiley & Sons, Inc.

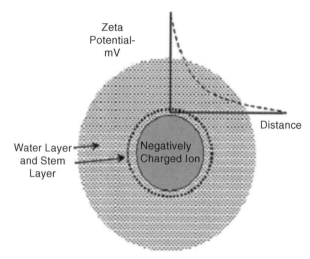

FIGURE 18.1 Zeta potential of a Colloid (Ionic charges and double layer around particles).

The measure of coagulant effects is Zeta potential. The Zeta potential is a measure of the electrochemical charge of a particle and the layer of surrounding ions of opposite charge. For example, if a particle has a negative charge, it will be surrounded with a layer of positively charged H^+ ions surrounding the particle. It is this layer of particles that helps make a colloid stable. The Zeta potential is measured by the mobility of colloidal particles across a cell. For many stable colloids in wastewater, the measured Zeta potential is between -16 mV and -22 mV but can range from -3 to -40 mV.[1] Coagulation generally occurs when the Zeta potential is lowered to less than ± 0.5 mV (Fig. 18.1).

Another way of describing coagulation is that it occurs when the surface charge is lowered enough to permit van der Waals attractive forces to make particles adhere when they collide or interact. Different valences of ions have varying effects in reducing the Zeta potential. The charge on the ion and the size of the ion also have an effect on the same. By comparison, for a specific coagulation effect, KCl takes 103 mg/l, K_2SO_4 requires 0.219 mg/l, and $K_3(FeCN)_6$ takes 0.096 mg/l to achieve the coagulation. Similar effects take place with cations (Fig. 18.2).

[1]Eckenfelder W. Industrial Water Pollution Control. New York: McGraw Hill, 1966.

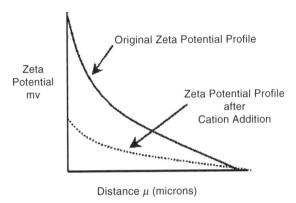

FIGURE 18.2   Effects of cations on Zeta potential.

Commercial Zeta potential units measure the net charge on ions, and the Zeta potential is often plotted on one axis with the turbidity of the sample on a parallel axis against the coagulant dose. The minimum turbidity is selected as the optimum point for precipitation and chemical dose.

Although good Zeta potential is used for measuring the charge for coagulation, it should never be the sole measure of determining coagulant dosage. That job is left to the jar test, and in fact, the Zeta potential is more of a confirmation of the observations of jar testing. The Zeta potential allows for optimization of dose, but then so does a jar test, without the expensive analyzer.

High-weight and high-charged molecular charge polymers are also used as coagulant aids. These are predominantly valuable because the equivalent charge is many hundred times that of even trivalent ions, and the effect of polymers can be substantial in reducing ion consumption. One milligram per liter of polymer added to a solution can replace as much as 30–50 mg/l of other salts.

Sols

Before the invention of polymers and their application to wastewater, sols were used as an early form of coagulant aid. Silica sols are semi-stable emulsions generally made from sodium silicate. The sodium silicate solution is highly alkaline, and it is diluted with water to a strength of several grams per liter, and then the solution is back-titrated to a near neutral pH with a combination of acid and cations or other anions including chlorine. Sols serve as a nucleating and weighting agent and can make a fragile floc substantially stronger and heavier. Sols are an inexpensive alternative to polymers. A brochure on the preparation of silica sols can be obtained from the Technical Service Division of Philadelphia Quarts (PQ Corp.) via e-mail.

FLOCCULATION AND MIXING

Flocculation is also known as mixing. It is a slow, thorough, and low shear mixing. Thomas Camp investigating the phenomenon back in the 1940s found that the average gradient G is significant in mixing.

$$G \text{ is the gradient} = (P/V\mu)^{0.5}$$

where G is in s^{-1}; P = power input in ft-lb/s (1 foot pound-force/second = 1.3558179 joule/second); V = mixing chamber volume in ft^3 (1 cubic foot = 28.316846712 cubic decimeter); μ = absolute viscosity of the fluid in $lbf\text{-}s/ft^2$ (1 pound-force second/square foot = 47,880.259 centipoise).

Fragile flocs such as biological flocs use $G = 10\text{--}30$
Medium strength (turbidity flocs) use $G = 20\text{--}50$
Chemical precipitation flocs use $G = 40\text{--}100$

For hydraulic mixing chambers use $P = Qwh$
where Q = flow rate in ft^3/s; w = weight of fluid in lb/ft^3; h = friction head loss.

Flocculators have slow mixing. Tanks with slowly rotating paddles, or other mixing devices, including baffles, and even air bubbles are adequate for the purpose. The most common type is a center shaft with opposing paddles set transverse to the length of the flow. A good discussion on the issue can be found in The Water Environment Federation Manual of Practice Number 8 (WEF MOP#8). Several older water treatment books also contain good drawings of flocculators.[2]

---

[2]WEF MOP#8 and Fair Gordon and Geyer FM. Elements of Water Supply and Wastewater Disposal. New York: John Wiley & Sons, 1950.

PRACTICE

The key to good flocculation is the jar test apparatus. It is a multigang paddle stirrer. It is best used with medium- to large-sized beakers of 600–1000 ml. After varying dosages of chemical addition, the paddle speed is turned up to give a flash mix and then turned way down to a very few revolutions per minute to promote floc growth. The best gauge is the visual formulation of the appearance of the floc and the clarity of the water. At the end of the flocculation, the paddle stirrer is removed and then the ability of the floc to settle and condense is examined.

Aliquots of the wastes are examined and further processed as may be necessary.

One note of caution on settling tests. Edge effects of the container can shape the performance of the material and provide false indications of the ability of a particular floc to behave in the desired manner. A minimum of 1 l should be used for this type of experiment, and large diameter vessels are better than small ones.

MODELING

There are at least three or four different theoretical models available for modeling coagulation. The models involve determining the particle sizes, the shear rate, dynamic viscosity of the fluid, and collisions per unit time. Unfortunately, most of these models require much more work to predict the results of a simple jar test and are useful only as research tools.

A recent work by a PhD student at the University of Ghent and a conference on Population Balance Modeling indicate the status of the work in the field and how much knowledge there is and how much is still to be gained.[3] The use of Computational Fluid Dynamics is being applied to flocculation theory, but it is still difficult to predict the size and distribution of sizes of various types of flocs, let alone the number of collisions, and when Zeta potential is introduced into the equations, the effort rapidly becomes a substantial research problem.

---

[3]Govorneau, Ruxandra. Activated Sludge Flocculation Dynamics: Online Measurement Methodology and Modeling [PhD Thesis], 2003–2004. Available at the Biomath Web site for the University of Ghent and Proceedings of the 2nd International Conference on Population Balance Modeling, Valencia, Spain, May 5–7, 2004.

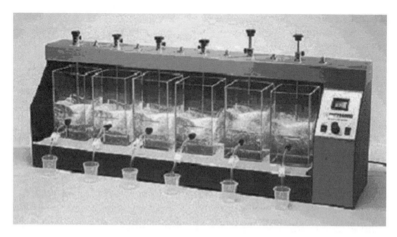

FIGURE 18.3   Multigang strirrer for jar testing (by Cole Palmer). http://www. coleparmer. com/catalog/catalog_images/large_images/9952100.jpg

The photograph in Figure 18.3 shows a six-port gang stirrer available from Cole Palmer and other laboratory supply companies for around $3000. The equipment is also available for lease.

19

WASTE TOPICS

Oily wastewaters
Blood and protein
Milk wastes
Refinery wastes
Metal plating wastes
Starch wastes
Phenols and chemical plant wastes
Small waste flows
Final thoughts

This chapter by design does not have a specific structure. Its sole purpose is to provide information, which does not fit anywhere else.

OILY WASTEWATERS

Treat oily wastewaters with high molecular weight polymers and high charge metal cations until the emulsion breaks. When the emulsion breaks, they can be treated with flotation or gravity settling for removal and collection of the oils.

There are two types of emulsions—physical and chemical. Physical emulsions are relatively easy to treat and break. Chemical emulsions such as water-soluble oils are more difficult until the oil has been worked or used in machining. Fresh chemical emulsions are extremely difficult to break.

For difficult emulsions, attack the emulsion either with aluminum sulfate or with ferric chloride until it breaks and then adjust your treatment. For extremely difficult emulsions, consider lowering the pH to less than 2 for

Practical Wastewater Treatment, by David L. Russell
Copyright © 2006 John Wiley & Sons, Inc.

about 10 min, add alum or ferric salts until the emulsion breaks, and then bring it slowly back with sodium hydroxide or lime. The liquid will probably have a slightly yellow cast indicating high dissolved salts content, but some of that can be reduced if ferric sulfate or aluminum sulfate (alum) is used and the system is neutralized with lime. The net effect is that the total dissolved solids in the system will be increased by about 1700 mg/l, which is the solubility of calcium sulfate. The alum/lime floc is usually very large and dense and settles quickly, scrubbing the water as it coagulates and settles. If this system is used, use sulfuric acid for pH adjustment, because it too will fall out with the lime precipitation. Other techniques using chemical precipitation can work as well.

Vegetable oils including palm, canula, and parrafins tend to be more difficult to de-emulsify. Successful treatments include cationic polymers followed by filtration at very low rates. The effluent is clear and free of suspended solids but may contain some dissolved oils.

The techniques described above also work on blood, proteins, and cooked starches in suspension. The acid treatment tends to coagulate the proteins and the starches. If the plant is planning on using the recovered wastewater starches as animal foods, then USDA-approved polymers must be used to avoid poisoning of the livestock (usually pigs). Raw starches do not precipitate well. The principal disadvantage of this technique is that the lime will cause a substantial increase in carbonate hardness of up to 40 grains.

BLOOD AND PROTEIN

It is very difficult to remove blood from the water. The best you could do is to remove the suspended blood and protein and to leave a very yellow solution behind. Chemical pretreatment lends itself well to low pH and lignin sulfonate. The sulfonate is a by-product of paper manufacturing and is generally readily available from local suppliers. The lignin sulfonate needs to be added at approximately the same rate or proportion each time. The recommended treatment is lowering the pH to below pH 4, followed by a consistent addition of lignin sulfonate at the same stoichiometric rates. Varying concentrations of blood in the water make the use of batch treatment and jar testing necessary.

Blood wastes are extremely high in BOD and nitrogen but poor in phosphorus. The waste can be treated by equalization and some pre-precipitation followed with high-rate anaerobic treatment. The wastewater is amenable to biological treatment, particularly anaerobic pretreatment followed by aerobic treatment. Phosphorus addition is a necessity. Chemical precipitation pretreatment can substantially reduce the amount of BOD, but there is still a substantial amount of dissolved BOD in the waste stream.

A relatively new process – the SHARON®-Anammox process – is being tested for high-strength ammonia wastes such as digester supernatant. The process was reported to have been successfully pilot tested for over 2 years and has been installed and operating in the 91st Street Wastewater Treatment Plant in Phoenix, Arizona.

MILK WASTES

These wastes are extremely deficient in nitrogen. Add ammonia and pretreat anaerobically followed by aerobic treatment, depending upon the strength of the waste. Without a nitrogen source, the milk wastes will not flocculate after biotreatment.

REFINERY WASTES

Start with as much pretreatment as you can afford. An API separator, unaided, is of limited use as a pretreatment device and tends to be more of a spill prevention control device. An API separator will remove oils down to about 15–50 mg/l depending upon the type of material, and for any emulsified product, there is no removal. Chemical coagulation combined with dissolved air flotation is definitely recommended.

An API separator is sometimes quite inadequate, and a coalescing filter or chevron type coagulation system can substantially enhance the performance of an API separator. For small flows and applications, consider using a prefilter and then a cartridge style coalescing filter—similar to that made by Serfilco, Pall, and others. These systems all tend to, work very well and remove free oils down to less than 15 mg/l.

Aerobic treatment—activated sludge works very well, but watch out for dumping of various chemicals, which can create toxicity problems, and conduct extensive pilot tests to determine long-term performance. Shock loading is also a significant problem. Equalization is strongly recommended. Extended aeration is also very good as a form of treatment for refinery wastes. The wastes may need a sewage or domestic waste source for nutrient balance and for some viable bacteria. Refinery wastes tend to be rich in carbon and poor in nitrogen and phosphorus. Vegetable oils behave differently than petroleum based oils.

METAL PLATING WASTES

You may need to use multiple stage precipitation. If there is hexavalent chromium, you will need to reduce it with sulfide (either H_2S or Na_2S) in an acid medium below pH 2 and then to neutralize it with lime to help bring down the chromium. Good removals are possible. Optimizing metals removal is often better done at two different pH values. In certain instances, effluent concentrations have been found to be less than 5 μg/l (5 ppb) after filtration.

If the waste contains oil, remove the oil first, and then handle the other waste problems.

STARCH WASTES

These occur in potato processing and other industries. Raw starch is difficult to treat. Neither does it filter well nor does it coagulate well. It is carbon rich and can be treated anaerobically. Cooked starch is substantially easier to treat. It will coagulate in acid followed by neutralization.

PHENOLS AND CHEMICAL PLANT WASTES

Phenols can be treated with acclimatized systems. The systems will have high-dilution and long-aeration periods. Acclimatized bacteria can be found in the soils around the plant and extracted, and in concentrations of up to 1500 mg/l, toxicity problems can be overcome and activated sludge treatment is recommended. Shock loading is a problem with this type of system.

SMALL WASTE FLOWS

For small and medium waste flows, consider the use of a Sequencing Batch Reactor (SBR). Do not let plant engineering try to make it a continuous reaction system (Activated Sludge). The SBR has come into its own, and it is an inexpensive method of treating low and high volume wastes. The advantage of the SBR is that it eliminates the need for the clarifier, with substantial savings in capital cost.

FINAL THOUGHTS

It is imperative that one has a complete understanding of the process before attempting to treat the wastewaters and developing a process solution. For many industries, improper characterization of the waste flows is the greatest cause of design failures. Shift change and cleanup shift where the process is not continuous are always the greatest generators of wastes and must be captured in the sampling data.

If there is an opportunity to do so, perform waste minimization and re-engineer the process to eliminate the waste stream, or reduce its volume before conducting sampling for a new design. The reduced volume and loading will often save enough money to pay for a portion of the treatment plant and the pilot plant study.

One of the greatest challenges to environmental control in a chemical plant is the idea that the operator has access to an "in-plant" sewer. This gives the operator a location to waste "out of specification" chemical batches and miscellaneous wastes, which either directly or eventually find their way into the wastewater treatment plant. This is a habit that needs to be broken. Sometimes it takes drastic action to reduce or eliminate the waste streams. In several instances, the best remedy to reduce or eliminate process wastes from a particular area was to plug the sewers and then observe the operators having to deal with the mess they were creating. While this is a hard lesson and requires plant management concurrence, it is always effective and has worked every time it was tried.

If one is attempting to design a wastewater treatment plant by the current U.S. design codes, it is well to remember that the codes have embodied a 30–50% safety factor into the design values, and that piling additional safety factors on top of that may be wasteful and unnecessary. Pilot plants and modeling are strongly recommended wherever it is possible.

INDEX

Practical Wastewater Treatment, by David L. Russell
Copyright © 2006 John Wiley & Sons, Inc.